PUBLISHER COMMENTARY

NASA LIFTING STANDARD
NASA-STD-8719.9B – Effective Date: 2018-10-25

This standard establishes NASA's minimum requirements for, but not all details of, the design, construction, testing, inspection, maintenance, operation, and personnel licensing of lifting devices and equipment (LDE) in order to enhance safety and reliability and ensure compliance with regulatory requirements. This standard is applicable to NASA-owned and NASA contractor-supplied overhead and gantry cranes (including top running, monorail, underhung, and jib cranes), mobile cranes, derricks, hoists, winches used for lifting applications, hoist-supported personnel lifting devices, load positioning devices, load measuring devices, hooks, jacks used for critical lifts, slings, rigging hardware, below-the-hook lifting devices, mobile aerial platforms, and high lift industrial trucks used in support of NASA operations at NASA installations and NASA operations in host countries. This standard does not apply to front-end loaders, elevators, or lifting devices used in non-lifting applications (e.g., balloon launching fixtures, jacks serving only to render casters ineffective).

Why buy a book you can download for free? We print this book so you don't have to.

If you find a good copy, you could print it using a network printer you share with 100 other people (typically its either out of paper or toner). If it's just a 10-page document, no problem, but if it's 250-pages, you will need to punch 3 holes in all those pages and put it in a 3-ring binder. Takes at least an hour. It's much more cost-effective to just order the latest version from Amazon.com. This book includes original commentary which is copyright material. Note that government documents are in the public domain. We print these large documents as a service so you don't have to. The books are compact, tightly-bound, full-size (8 ½ by 11 inches), with large text and glossy covers. 4th Watch Publishing Co. is a HUBZONE SDVOSB.
https://usgovpub.com www.usgovpub.com

Copyright © 2018 4th Watch Publishing Co. All Rights Reserved

List of Applicable Publications:

NASA RCM	NASA Reliability-Centered Maintenance (RCM)
NASA	Systems Engineering Handbook (SP-2016-6105)
Probabilistic Risk Assessment Procedures Guide for NASA Managers and Practitioners	
UFC 4-010-06	Cybersecurity of Facility-Related Control Systems
NIST SP 800-82	Guide to Industrial Control Systems (ICS) Security
Whitepaper	NIST Framework for Improving Critical Infrastructure Cybersecurity
NISTIR 8170	The Cybersecurity Framework
FC 4-141-05N	Navy and Marine Corps Industrial Control Systems Monitoring Stations
UFC 3-430-11	Boiler Control Systems
NISTIR 8089	An Industrial Control System Cybersecurity Performance Testbed
NIST SP 800-12	An Introduction to Information Security
NIST SP 800-18	Developing Security Plans for Federal Information Systems
NIST SP 800-31	Intrusion Detection Systems
NIST SP 800-34	Contingency Planning Guide for Federal Information Systems
NIST SP 800-35	Guide to Information Technology Security Services
NIST SP 800-39	Managing Information Security Risk
NIST SP 800-40	Guide to Enterprise Patch Management Technologies
NIST SP 800-41	Guidelines on Firewalls and Firewall Policy
NIST SP 800-44	Guidelines on Securing Public Web Servers
NIST SP 800-47	Security Guide for Interconnecting Information Technology Systems
NIST SP 800-48	Guide to Securing Legacy IEEE 802.11 Wireless Networks
NIST SP 800-53A	Assessing Security and Privacy Controls
NIST SP 800-61	Computer Security Incident Handling Guide
NIST SP 800-77	Guide to IPsec VPNs
NIST SP 800-83	Guide to Malware Incident Prevention and Handling for Desktops and Laptops
NIST SP 800-92	Guide to Computer Security Log Management
NIST SP 800-94	Guide to Intrusion Detection and Prevention Systems (IDPS)
NIST SP 800-97	Establishing Wireless Robust Security Networks: A Guide to IEEE 802.11i
NIST SP 800-137	Information Security Continuous Monitoring (ISCM)
NIST SP 800-160	Systems Security Engineering
NIST SP 800-171	Protecting Controlled Unclassified Information in Nonfederal Systems
NIST SP 1800-7	Situational Awareness for Electric Utilities
NISTIR 7628	Guidelines for Smart Grid Cybersecurity
DoD	Energy Manager's Handbook
FEMP	Operations & Maintenance Best Practices
UFC 4-020-01	DoD Security Engineering Facilities Planning Manual
UFC 4-021-02	Electronic Security Systems by Department of Defense
GSA	GSA Courtroom Technology Manual
Draft NISTIR 8179	Criticality Analysis Process Model
NISTIR 8144	Assessing Threats to Mobile Devices & Infrastructure
NISTIR 8151	Dramatically Reducing Software Vulnerabilities
NIST SP 800-183	Networks of 'Things'
NIST SP 800-184	Guide for Cybersecurity Event Recovery

Copyright © 2018 4th Watch Publishing Co. All Rights Reserve

MEASUREMENT SYSTEM
IDENTIFICATION

NASA-STD-8719.9B

Approved: 2018-10-25
Superseding NASA-STD-8719.9A

LIFTING STANDARD
NASA TECHNICAL STANDARD

National Aeronautics and Space Administration

NASA-STD-8719.9B – 2018-10-25

DOCUMENT HISTORY LOG

Status	Document Revision	Approval Date	Description
Baseline		05-09-2002	Conversion of document to NASA-STD format. Addition of sections on mobile aerial platforms, powered industrial trucks, and jacks. Addition of appendices on lifting personnel with a crane and using a crane to load test other lifting equipment. Designation of an installation Lifting Devices and Equipment Manager (LDEM) is also required.
Change	1	10-01-2007	Document revalidation without changes other than updates to Cover, Foreword (address), and Revision Log.
Change	2	09-28-2012	Document revalidation without changes other than updates to Cover, Foreword (address), and Revision Log.
Revision	A	08-13-2015	Added a General LDE Requirements chapter containing requirements common to all LDE. Deleted repetitive requirements from OSHA and NCS throughout the document. Incorporated applicable NCS by reference throughout the document. Added new appendices referencing critical lift requirements and LDEM roles, approvals, and special permissions. Removed appendices that were duplicative of OSHA and NCS.
Revision	B	10-25-2018	Global: replaced "structural sling" with "below-the-hook lifting device". Global: replaced "NCS" with "VCS". Global: replaced "National Consensus Standard" with "Voluntary Consensus Standard". Global: minor punctuation, formatting, and editorial changes. Foreword: removed text regarding changes made in previous revision. 1.3.7 Added "Center/Facility" to clarify level of responsible organizations. 2.1 Updated the Applicable Documents General information. 2.1.2 Added document reference "ANSI/ITSDF B56.14-2015 Safety Standard for Vehicle Mounted Forklift Trucks. 3.1 Deleted acronym definition for "SARD" (not used); replaced "NCS" with "VCS". 3.2 Added definitions for "attachment", "below-the-hook-lifting device", "non-load test slings, rigging hardware, and below-the-hook lifting devices", and "periodic inspection"; modified "critical lift" definition to remove limit of 75% capacity on mobile cranes and clarify; deleted "lifting equipment", and

"wire rope sling" definitions and incorporated in "lifting devices and equipment"; modified "lifting device", "load", "periodic load test", "rigging hardware", "sling", "structural sling" and configuration management definitions; modified "nondestructive testing" definition to delete "development and", "replaced "national consensus standard" with "voluntary consensus standard" with the same definition, and deleted the definitions "can", "may, "shall", and "should".

4.2 Clarified responsibilities re: Critical lifts; deleted 75% limit on mobile cranes for critical lifts.

4.3.1 Added requirement for LDEM approval of safety hazard analysis of critical lift/custom LDE, with exclusion.

4.4.2 Added exclusion to requirement for LDEM approval of certain critical lift LDE.

4.5.2 Added note clarifying that periodic load tests are at a lower load than proof tests.

4.5.5 Clarified applicability of notes.

4.7.7 Added note for emphasis re: OSHA rqmt.

4.7.10 Reinstated rqmt regarding assessing lifts of energetic materials, and reference to 8719.12.

4.9.2, 4.9.3, 4.9.4 Clarified marking requirements in notes.

4.11.2.3 Added reference to NPR 1800.1.

5.7.3 Clarified rqmts for use of OH cranes for load testing other LDE, allowed exceeding 50% of crane rating in certain cases.

5.9.2 Reinstated requirement for OH crane directional labeling.

6.7.2 Added section regarding ascertaining load weight in light of removal of limit on mobile crane critical lifts to 75% of crane rating.

6.7.7 Deleted redundant statement regarding use of operational aids.

7.4.1.1b Reworded for clarity and emphasis.

7.4.2.1, 7.4.2.5, 7.4.2.6 Allowed for use of hoist/winch with single upper limit and no lower limit switch for critical lifts with certain provisions.

7.7.8 Rewritten to allow exceeding 50% capacity when performing periodic load tests of items when items are freely hanging from the hook.

10.1.2 10.5 Clarified that high lift industrial truck attachments (as well as the truck) must comply with standards and mfr recommendations, and proof and periodic load test requirements.

10.3 Allowed exception to safety hazard analysis with LDEM approval.

10.5.2.5 Added requirement for load test of forklift attachments.

| | | | 10.6.2 Expanded the list of items on a forklift requiring annual inspection.
10.9 Clarified that OSHA and VCS require marking of industrial trucks to identify approved attachments.
13.6.3 Clarified NDT requirements for hooks.
14 Replaced "structural slings" with "below-the-hook lifting devices" throughout.
14.5.2.2 Add to note a sentence regarding who is manufacturer of an assembly with respect to determining testing.
14.5.2.3 Clarify meaning of "permanently attached".
14.5.2.4 Corrected proof load test requirement for consistency with VCS.
14.5.3.4 Reinstated provision exempting permanently attached hardware from periodic load test.
Appendix B Updated to reflect changes made elsewhere regarding critical lifts.
Appendix C Updated to reflect changes made elsewhere regarding LDEM roles, approvals, and special permissions.
Appendix D added to provide recommended minimum LDEM qualifications. |
|---|---|---|---|
| | | | |

A note concerning the history of this document:

The original NASA Safety Standard for Lifting Devices and Equipment was issued as NSS/GO-1740.9 in July 1982. In July 1988 it was revised, and Revision A was issued reflecting significant changes related to mobile cranes, hoist-supported personnel platforms, personnel lifting buckets, and guidance concerning super critical lifts. In November 1991 it was revised again, and Revision B was issued which deleted the guidance on super critical lifts and added the NASA Alternate Standard for Suspended Load Operations. Additional revisions were issued as change pages in March 1993 to expand operational test requirements for hoist-supported personnel lifting devices. When the time came to update the standard again, in addition to the technical changes to the document (synopsized in the Revision Log above), the format and numbering were changed to reflect current practices and conventions for NASA Standards.

NASA-STD-8719.9B – 2018-10-25

FOREWORD

This NASA Technical Standard provides Agency-level requirements for the design, construction, testing, inspection, maintenance, operation, and personnel licensing requirements for lifting devices and equipment used in support of NASA operations. With the exception of the NASA Alternate Standard for Suspended Load Operations contained in Appendix A, this standard is not inclusive of, or a substitute for, Occupational Safety and Health Administration (OSHA) or additional government regulations (including applicable host country regulations). This standard provides NASA-specific requirements and references applicable OSHA and Voluntary Consensus Standards (VCS).

This document establishes minimum safety requirements; NASA installations are encouraged to assess their individual lifting programs and develop additional requirements as needed.

This standard is approved for use by NASA Headquarters and NASA Centers, including Component Facilities, and is intended to be applied on NASA contracts. It may also apply to the Jet Propulsion Laboratory and other contractors only to the extent specified or referenced in applicable contracts.

This standard was developed by the NASA Office of Safety and Mission Assurance (OSMA). Requests for information, corrections, or additions to this standard should be submitted to the OSMA by email to Agency-SMA-Policy-Feedback@mail.nasa.gov or via the "Email Feedback" link at https://standards.nasa.gov.

_____
Terrence W. Wilcutt
Chief, Safety and Mission Assurance

10/25/2018
Approval Date

NASA-STD-8719.9B – 2018-10-25

TABLE OF CONTENTS

SECTION	PAGE
DOCUMENT HISTORY LOG	2
1. SCOPE	11
1.1 Purpose	11
1.2 Applicability	11
1.3 Roles and Responsibilities	12
1.4 Order of Precedence	13
1.5 Requests for Relief	13
1.6 Using this Standard	14
2. APPLICABLE DOCUMENTS	15
2.1 General	15
3. DEFINITIONS AND ACRONYMS	18
3.1 Abbreviations and Acronyms	18
3.2 Definitions	19
4. GENERAL LDE REQUIREMENTS	26
4.1 General	26
4.2 Classification of Lifts	26
4.3 Safety Hazard Analysis	27
4.4 Design	27
4.5 Testing	27
4.6 Inspection	28
4.7 Operation	29
4.8 Maintenance	30
4.9 Labeling and Tagging	30
4.10 Records	31
4.11 Personnel Training and Licensing	31
5. OVERHEAD CRANES	33
5.1 General	33
5.2 Classification of Lifts	34
5.3 Safety Hazard Analysis	34
5.4 Design	34
5.5 Testing	36
5.6 Inspection	38

NASA-STD-8719.9B – 2018-10-25

TABLE OF CONTENTS

| **SECTION** | **PAGE** |

5.7	Operation	38
5.8	Maintenance	40
5.9	Labeling and Tagging	40
5.10	Records	40
5.11	Personnel Training and Licensing	41

6. MOBILE CRANES AND DERRICKS 41
- 6.1 General 41
- 6.2 Classification of Lifts 41
- 6.3 Safety Hazard Analysis 41
- 6.4 Design 41
- 6.5 Testing 42
- 6.6 Inspection 44
- 6.7 Operation 44
- 6.8 Maintenance 46
- 6.9 Labeling and Tagging 46
- 6.10 Records 46
- 6.11 Personnel Training and Licensing 46

7. HOISTS AND WINCHES 46
- 7.1 General 46
- 7.2 Classification of Lifts 47
- 7.3 Safety Hazard Analysis 47
- 7.4 Design 47
- 7.5 Testing 50
- 7.6 Inspection 53
- 7.7 Operation 53
- 7.8 Maintenance 55
- 7.9 Labeling and Tagging 55
- 7.10 Records 55
- 7.11 Personnel Training and Licensing 55

8. HOIST-SUPPORTED PERSONNEL LIFTING DEVICES 56
- 8.1 General 56
- 8.2 Classification of Lifts 56

TABLE OF CONTENTS

SECTION **PAGE**

    8.3   Safety Hazard Analysis 56
    8.4   Design 57
    8.5   Testing 57
    8.6   Inspection 59
    8.7   Operation 60
    8.8   Maintenance 60
    8.9   Labeling and Tagging 60
    8.10   Records 60
    8.11   Personnel Training and Licensing 61

9. MOBILE AERIAL PLATFORMS 61
    9.1   General 61
    9.2   Classification of Lifts 61
    9.3   Safety Hazard Analysis 61
    9.4   Design 61
    9.5   Testing 62
    9.6   Inspection 63
    9.7   Operation 63
    9.8   Maintenance 63
    9.9   Labeling and Tagging 63
    9.10   Records 63
    9.11   Personnel Training and Licensing 63

10. HIGH LIFT INDUSTRIAL TRUCKS 64
    10.1   General 64
    10.2   Classification of Lifts 64
    10.3   Safety Hazard Analysis 64
    10.4   Design 64
    10.5   Testing 65
    10.6   Inspection 66
    10.7   Operation 67
    10.8   Maintenance 67
    10.9   Labeling and Tagging 67
    10.10   Records 67
    10.11   Personnel Training and Licensing 68

NASA-STD-8719.9B – 2018-10-25

TABLE OF CONTENTS

SECTION	PAGE

11. LOAD POSITIONING AND LOAD MEASURING DEVICES 68
- 11.1 General 68
- 11.2 Classification of Lifts 68
- 11.3 Safety Hazard Analysis 68
- 11.4 Design 69
- 11.5 Testing 69
- 11.6 Inspection 70
- 11.7 Operation 71
- 11.8 Maintenance 71
- 11.9 Labeling and Tagging 71
- 11.10 Records 71
- 11.11 Personnel Training and Licensing 71

12. JACKS 72
- 12.1 General 72
- 12.2 Classification of Lifts 72
- 12.3 Safety Hazard Analysis 72
- 12.4 Design 73
- 12.5 Testing 73
- 12.6 Inspection 74
- 12.7 Operation 75
- 12.8 Maintenance 75
- 12.9 Labeling and Tagging 75
- 12.10 Records 75
- 12.11 Personnel Training and Licensing 75

13. HOOKS 75
- 13.1 General 75
- 13.2 Classification of Lifts 76
- 13.3 Safety Hazard Analysis 76
- 13.4 Design 76
- 13.5 Testing 76
- 13.6 Inspection 77
- 13.7 Operation 77

TABLE OF CONTENTS

SECTION	PAGE
13.8 Maintenance	77
13.9 Labeling and Tagging	78
13.10 Records	78
13.11 Personnel Training and Licensing	78
14. SLINGS, RIGGING HARDWARE, AND BELOW-THE-HOOK LIFTING DEVICES	78
14.1 General	78
14.2 Classification of Lifts	79
14.3 Safety Hazard Analysis	79
14.4 Design	79
14.5 Testing	79
14.6 Inspection	82
14.7 Operation	82
14.8 Maintenance	83
14.9 Labeling and Tagging	83
14.10 Records	83
14.11 Personnel Training and Licensing	84
APPENDIX A: NATIONAL AERONAUTICS AND SPACE ADMINISTRATION ALTERNATE STANDARD FOR SUSPENDED LOAD OPERATIONS	85
APPENDIX B: SUMMARY OF CRITICAL LIFT REQUIREMENTS	93
APPENDIX C: LDEM ROLES, APPROVALS, AND SPECIAL PERMISSIONS	99
APPENDIX D: RECOMMENDED MINIMUM LDEM QUALIFICATIONS	104

NASA-STD-8719.9B – 2018-10-25

Lifting Standard

1. SCOPE

1.1 Purpose

1.1.1 This standard establishes NASA's minimum requirements for, but not all details of, the design, construction, testing, inspection, maintenance, operation, and personnel licensing of lifting devices and equipment (LDE) listed in section 1.2 in order to enhance safety and reliability and ensure compliance with regulatory requirements.

1.1.2 This standard provides NASA-specific requirements and references applicable OSHA and Voluntary Consensus Standards (VCS). This standard establishes minimum safety requirements; NASA installations are encouraged to assess their individual lifting programs and develop additional requirements as needed.

1.2 Applicability

1.2.1

1.2.2 Rented or leased LDE used for non-critical lifts may be exempted from this standard by the written decision of the contracting officer, the responsible NASA installation/program safety office, and the Lifting Devices and Equipment Manager (LDEM), based on an assessment of associated risk.

1.2.3 The need for compliance with this standard at contractor installations performing NASA work should be evaluated and made a contractual requirement where deemed necessary by the contracting officer, the responsible NASA installation/program safety office, and the LDEM.

1.2.4 The LDEM shall have the authority to interpret this standard.

1.2.5   The LDEM shall have the authority to approve, disapprove, and levy requirements for the use of LDE not covered by paragraph 1.2.1 due to safety concerns or hazards presented by a particular application.

1.2.6   In this standard, "shall" denotes a mandatory requirement, "may" denotes a discretionary privilege or permission, "can" denotes statements of possibility or capability, "should" denotes a good practice, "will" denotes an expected outcome, and "must" denotes a reference to an existing requirement.

1.3   Roles and Responsibilities

1.3.1   The Chief, Safety and Mission Assurance assures that NASA Centers, Component Facilities, and programs protect personnel and property from the hazards posed by LDE in accordance with this standard.

1.3.2   The Director, Safety and Assurance Requirements Division establishes the NASA Lifting Devices and Equipment Committee (LDEC), with membership to include the Center/Facility LDEMs, and designates the LDEC chair.

1.3.3   The NASA LDEC reviews proposed changes to this standard and serves as a forum for the exchange of LDE information and issue resolution.

1.3.4   The Center Director of each NASA Center and Component/Facility shall:

   a. Designate in writing one person as the LDEM to perform the duties delineated in paragraph 1.3.6 and at least one person as the Alternate LDEM to assist the LDEM.

   b. Provide adequate resources to implement the requirements of this standard.

   c. Establish a Center/Facility LDEC.

1.3.5   The Center/Facility LDEC shall review Center/Facility-level LDE safety policy and requirements and serve as a forum for the exchange of LDE information and issue resolution.

1.3.6   The LDEM shall perform the following activities:

   a. Participate as a member of the NASA LDEC.

   b. Chair the Center/Facility LDEC.

   c. Serve as the focal point for implementation, clarification, and enforcement of this standard at the Center/Facility.

1.3.7   The Center/Facility organizations responsible for LDE shall ensure the following activities are performed:

a. In coordination with the LDEM, ensure LDE is designed, constructed, tested, inspected, maintained, and operated in accordance with this standard.

b. In coordination with the LDEM, ensure personnel are trained and licensed in accordance with this standard.

c. Provide representation for the Center/Facility LDEC, with membership to include representatives of the responsible organizations for LDE.

1.4 Order of Precedence

1.4.1 This standard does not supersede any higher level safety requirements (such as OSHA regulations).

Note: This document supplements and provides implementation direction for OSHA regulations. With the exception of Appendix A, NASA Alternate Standard for Suspended Load Operations (which is approved by OSHA), it is not a substitute for any OSHA regulation. OSHA regulations stated in the Code of Federal Regulations (CFR) are law and, as such, apply to all NASA operations. Some states have their own OSHA programs which may apply additional, more stringent regulatory requirements.

1.4.2 VCS are mandatory when required by OSHA regulations or when required by this document as specified herein.

1.4.3 This document takes precedence over VCS except in those cases in which the VCS is invoked by regulation.

1.4.4 All document citations refer to the versions specified herein.

1.5 Requests for Relief

1.5.1 If a requirement cannot be met, all requests for relief to Agency-level Safety and Mission Assurance requirements shall be in accordance with NPR 8715.3 and NASA-STD-8709.20.

Note: The NASA request for relief process does not apply to Federal and applicable State/local regulations (e.g., OSHA, Cal/OSHA). Any relief to a Federal or State/local regulation must first be approved by the Chief, Safety and Mission Assurance (Headquarters Office of Safety and Mission Assurance), in accordance with NPR 8715.3, and then by the appropriate Federal/State/local agency (e.g., NASA Alternate Safety Standard for Suspended Load Operations approved by OSHA).

1.6 Using this Standard

1.6.1 This standard provides the minimum NASA requirements for the design, testing, inspection, maintenance, personnel licensing, and operation of LDE. It is not a comprehensive list of all applicable requirements.

1.6.2 The following steps should be taken to identify all requirements for a particular type of LDE:

a. Address applicable Federal regulations (e.g., CFR, OSHA).

Note 1: As previously noted, this document supplements and provides implementation direction for OSHA regulations. With the exception of Appendix A, NASA Alternate Standard for Suspended Load Operations (which is approved by OSHA), this standard is not a substitute for any OSHA regulation.

Note 2: OSHA regulations are available online at www.OSHA.gov.

b. Address applicable State and Local regulations (e.g., Cal/OSHA).

Note: Some state regulations may be more stringent than NASA requirements.

c. Address the general LDE requirements in Chapter 4 of this standard.

d. Address the LDE-specific requirements, as applicable, in Chapters 5-14 of this standard.

e. Address applicable VCS incorporated by reference in this standard.

Note: References to applicable VCS are provided throughout this standard. Most VCS are available online at: https://standards.nasa.gov/

f. Address any applicable Center-level LDE requirements.

Note: The requirements of this standard are the minimum requirements. Additional or more stringent LDE requirements may be developed at each Center.

1.6.3 In case of questions regarding conflicting requirements, the applicability of this standard, or to request a clarification, contact the LDEM.

1.6.4 Appendix B references the requirements that apply to critical lifts contained throughout this document. This appendix may be used to locate applicable critical lift requirements.

1.6.5 Appendix C references the instances within this standard where the LDEM has direct involvement in a particular activity. This appendix may be used to locate the applicable requirements.

2. APPLICABLE DOCUMENTS

2.1 General

The documents listed in this section contain provisions that constitute requirements of this standard as cited in the text. Use of more recent issues of cited documents may be authorized by the responsible LDEM.

2.1.1 Government Documents

29 CFR 1910	Occupational Safety and Health Standards
29 CFR 1926	Safety and Health Regulations for Construction
29 CFR 1960	Basic Program Elements For Federal Employee Occupational Safety and Health Programs and Related Matters
NPR 1800.1	NASA Occupational Health Program Procedures
NPR 8715.1	NASA Occupational Safety and Health Programs
NPR 8715.3	NASA General Safety Program Requirements
NPR 8715.7	Expendable Launch Vehicle (ELV) Payload Safety Program
NASA-STD-8709.20	Management of Safety and Mission Assurance Technical Authority (SMA TA) Requirements
NASA-STD-8719.24	NASA Expendable Launch Vehicle Payload Safety Requirements
NRRS 1441.1	NASA Records Retention Schedules

2.1.2 Non-Government Documents

ANSI/ITSDF B56.1-2012	Safety Standard for Low Lift and High Lift Trucks
ANSI/ITSDF B56.6-2011	Safety Standard for Rough Terrain Forklift Trucks
ANSI/ITSDF B56.10-2012	Safety Standard for Manually Propelled High Lift Industrial Trucks
ANSI /ITSDF B56.14-2015	Safety Standard for Vehicle Mounted Forklift Trucks

ANSI/SAIA A92.2-2009	Vehicle Mounted Elevating and Rotating Aerial Devices
ANSI/SAIA A92.3-2006	Manually Propelled Elevating Aerial Platforms
ANSI/SAIA A92.5-2006	Boom Supported Elevating Work Platforms
ANSI/SAIA A92.6-2006	Self-Propelled Elevating Work Platforms
ASME B30.1-2009	Jacks, Industrial Rollers, Air Casters, and Hydraulic Gantries
ASME B30.2-2011	Overhead and Gantry Cranes (Top Running Bridge, Single or Multiple Girder, Top Running Trolley Hoist)
ASME B30.3-2012	Tower Cranes
ASME B30.4-2010	Portal and Pedestal Cranes
ASME B30.5-2014	Mobile and Locomotive Cranes
ASME B30.6-2010	Derricks
ASME B30.7-2011	Winches
ASME B30.8-2010	Floating Cranes and Floating Derricks
ASME B30.9-2014	Slings
ASME B30.10-2014	Hooks
ASME B30.11-2010	Monorails and Underhung Cranes
ASME B30.12-2011	Handling Loads Suspended From Rotorcraft
ASME B30.13-2011	Storage/Retrieval (S/R) Machines and Associated Equipment
ASME B30.14-2010	Side Boom Tractors
ASME B30.16-2012	Overhead Hoists
ASME B30.17-2006	Overhead and Gantry Cranes (Top Running Bridge, Single Girder, Underhung Hoist)
ASME B30.19-2011	Cableways
ASME B30.20-2013	Below-the-Hook Lifting Devices

ASME B30.21-2014	Lever Hoists
ASME B30.22-2010	Articulating Boom Cranes
ASME B30.23-2011	Personnel Lifting Systems
ASME B30.24-2013	Container Cranes
ASME B30.25-2013	Scrap and Material Handlers
ASME B30.26-2010	Rigging Hardware
ASME B30.28-2010	Balance Lifting Units
ASME B30.29-2012	Self-Erect Tower Cranes
ASME BTH-1-2014	Design of Below-the-Hook Lifting Devices
ASME HST-1-2012	Performance Standard for Electric Chain Hoists
ASME HST-2-2014	Performance Standard for Hand Chain Manually Operated Chain Hoists
ASME HST-3-1999	Performance Standard for Manually Lever Operated Chain Hoists
ASME HST-4-1999	Performance Standard for Overhead Electric Wire Rope Hoists
ASME HST-5-2014	Performance Standard for Air Chain Hoists
ASME HST-6-1999	Performance Standard for Air Wire Rope Hoists
CMAA Specification No. 70-2010	Specifications for Electric Overhead Traveling Cranes
CMAA Specification No. 74-2010	Specification for Top Running and Under Running Single Girder Electric Overhead Traveling Cranes
CP-189-2011	Qualification and Certification of Nondestructive Testing Personnel
DIN EN 13000-2014	Cranes-Mobile Cranes
NAS-410-2014	Certification & Qualification of Nondestructive Test Personnel
NFPA 70-2014	National Electric Code

SAE J1063-2013	Crane Structures, Cantilevered Boom, Method of Test
SAE J765-1990	Crane Load Stability Test Code
SNT-TC-1A-2011	Personnel Qualification and Certification in Nondestructive Testing
WSTDA-RS-1-2010	Recommended Standard for Synthetic Polyester Round Slings
WSTDA-WS-1-2015	Recommended Standard for Synthetic Web Slings
	Wire Rope User's Manual-4th Edition
	Wire Rope Sling User's Manual-3rd Edition

3. DEFINITIONS AND ACRONYMS

3.1 Abbreviations and Acronyms

AGMA	American Gear Manufacturers Association
ANSI	American National Standards Institute
ASME	American Society of Mechanical Engineers
ASNT	American Society for Nondestructive Testing
Cal/OSHA	California Occupational Safety and Health Administration
CEN	European Committee for Standardization
CFR	Code of Federal Regulations
CMAA	Crane Manufacturers Association of America
COTS	Commercial Off-The-Shelf
DIN	Deutsches Institut für Normung
E-Stop	Emergency Stop
FMEA	Failure Modes and Effects Analysis
ITSDF	Industrial Truck Standards Development Foundation
LDE	Lifting Devices and Equipment

LDEC	Lifting Devices and Equipment Committee
LDEM	Lifting Devices and Equipment Manager
NASA	National Aeronautics and Space Administration
NDE	Nondestructive Evaluation
NDT	Nondestructive Testing
NFPA	National Fire Protection Association
NPR	NASA Procedural Requirements
NSS/GO	NASA Safety Standard/Ground Operations
OEM	Original Equipment Manufacturer
OSHA	Occupational Safety and Health Administration
SAE	SAE International
SAIA	Scaffold and Access Industry Association
SFP	Single Failure Point
STD	Standard
VCS	Voluntary Consensus Standard
WRTB	Wire Rope Technical Board
WSTDA	Web Sling and Tie Down Association

3.2 Definitions

Attachment (for Industrial Trucks): Devices other than conventional forks or load backrest extensions, mounted permanently or temporarily on the elevating mechanism of an industrial truck for handling the load. Common types of attachments include but are not limited to fork extensions, clamps, rotating devices, side shifters, and booms.

Below-the-Hook Lifting Device: A device used for attaching a load to a hoist or other lifting mechanism. The device may consist of or contain components such as slings, hooks, and rigging hardware that are addressed by ASME B30 volumes or other standards. Common types include spreader bars, beam clamps, barrel lifters, and vacuum lifts. Some of these devices may be referred to as structural slings.

Brake: A device used for retarding or stopping motion.

Certified Equipment: Lifting device or equipment documented by the LDEM as complying with the design, construction, maintenance, test, and other requirements of this standard.

Configuration Management: Process for establishing and maintaining consistency of a product's functional and physical characteristics, evaluating and authorizing any changes to those characteristics, and recording and documenting the characteristics and any changes to them to verify compliance with the product's configuration requirements throughout its life.

Crane: A machine for lifting and lowering a load and moving it horizontally, with the hoisting mechanism an integral part of the machine.

Critical LDE: Lifting Devices and Equipment used to perform Critical Lifts.

Critical Lift: Lifts during which failure/loss of control presents an elevated risk of serious injury, loss of life, or loss of one-of-a-kind articles, high dollar items or major facility components whose loss would have serious programmatic or institutional impact. Lifts of high-value flight hardware and/or non-routine lifts (e.g., lift point below center of gravity) are usually classified as critical lifts, while lifts of small, improvised mini-satellites, for example, most likely would not be. Lifting and movement of flight hardware components packaged per applicable shipment specifications are typically not classified as critical lifts.

Derrick: An apparatus consisting of a mast or equivalent member held at the end by guys or braces, with or without a boom, for use with a hoisting mechanism and operating ropes.

Design Factor: A numeric usually expressed as a ratio of the ultimate strength or yield strength to the rated capacity. It is used in calculations to account for variations found in the properties of materials, manufacturing tolerances, operating conditions, and design assumptions.

Designated Person: A person who is qualified and who has been selected or assigned (in writing) by the responsible organization to perform specific duties.

Dummy Load: A test load used to simulate the real load; typically a test weight.

Dummy Rated Load: A test load equal to the rated load of the device; typically a test weight.

Eddy Current Brake: An electrical induction brake used to reduce or control speed.

Emergency Stop (E-Stop): A manually operated switch or valve to cut off electric or fluid power independently of the regular operating controls.

Equivalent Entity: A person or organization (including an employer) which by possession of equipment, technical knowledge and skills, can perform with equal competence the same repairs and tests as the person or organization with which it is equated.

Failure Modes and Effects Analysis (FMEA): A systematic, methodical analysis performed to identify and document failure modes and their resultant effects at a prescribed level.

Hazard: Any real or potential condition that can cause injury or death to personnel or damage to or loss of equipment or property.

Hoist: A machinery unit device used for lifting and lowering a load.

Hoist-Supported Personnel Lifting Device: Device specifically designed to lift and lower persons via a hoist. These devices include hoist-supported platforms where personnel occupy the platform during movement. These devices do not include elevators, lifting personnel with a crane, mobile aerial platform, or platforms hoisted unoccupied to a position and anchored or restrained to a stationary structure before personnel occupy the platform (refer to Personnel Access Platform).

Holding Brake: A brake that automatically prevents motion when power is off.

Idle Lifting Device: Lifting device that has not been used for 12 months or more, or that has no projected use for the next 12 months.

Jack: A mechanism with a base and load point designed for controlled linear movement.

Licensed Operator: See Licensed Personnel.

Licensed Personnel: Individuals documented by the LDEM as meeting the personnel licensing requirements of this standard. Licensed personnel may be referred to as certified personnel or certified operators in other regulations and VCS.

Lifting Device: A generic term or modifier broadly used to refer to both equipment that actively lifts (cranes, powered industrial trucks, etc.) and individual pieces or assemblies of components used in the lifting process (slings, hoist-supported lifting devices, shackles, etc.).

Lifting Devices and Equipment (LDE): Devices, equipment, and their accessories used to lift, lower, and position a load.

Lifting Devices and Equipment Manager (LDEM): Person designated by the Center Director, responsible for managing the installation lifting devices and equipment program, coordinating with appropriate personnel at their installation on lifting issues, and providing their installation's position on lifting devices and equipment safety issues.

Load: The total weight of the items being supported, raised, or moved by a lifting device or equipment, including rigging hardware, slings, below-the-hook lifting devices, the load block for some mobile crane configurations, or any other attachments that are not taken into account when determining the rated capacity of the lifting device or equipment.

Load Brake: A braking device that retards and controls the load during lowering and keeps the load from falling if the holding brake fails.

Load Measuring Device: A device below the hook, which is used to indicate the weight of the item being lifted.

Load Positioning Device: Instrument installed between the hook and load to allow precise control of lifting operations (e.g., Hydra Sets®).

Mobile Aerial Platform: A mobile device that has an adjustable position platform and is supported from ground level by a structure.

NASA Operation: Any activity or process under NASA direct control or that includes major NASA involvement.

Noncritical Lift: A lift involving routine lifting operations governed by standard industry rules and practices except as supplemented with unique NASA testing, operations, maintenance, inspection, and personnel licensing requirements contained in this standard.

Nondestructive Evaluation (NDE): See Nondestructive Testing.

Nondestructive Testing (NDT): The application of technical methods to examine materials or components in ways that do not impair future usefulness and serviceability in order to detect, locate, measure, and evaluate flaws; to assess integrity, properties, and composition; and to measure geometrical characteristics.

Non-Load Test Slings, Rigging Hardware, And Below-The-Hook Lifting Devices: Slings, rigging hardware, and below-the hook lifting devices meeting the criteria of section 14.5.4 and designated and approved by the LDEM as not subject to periodic load testing requirements.

Periodic Inspection: A thorough examination of LDE conducted at predetermined intervals (typically monthly to yearly) to assess the condition of the equipment. These inspections do not include pre-use inspections performed each day before the equipment is used. Details of these inspections are provided in regulations and industry standards.

Periodic Load Test: A load test performed at predetermined intervals to determine whether the equipment (e.g., limit switches, E-Stop, controls, brakes, slings, shackles) is functioning properly.

Personnel Access Platform: A platform, typically deployed or relocated by one or multiple dedicated hoists or winches, which allow personnel to access and work in a specific area of a fixed structure or building. Personnel occupy these platforms only after the platforms are deployed and secured and never during movement or while the platforms are supported by hoists/winches. For platforms specifically designed to lift and lower persons via a hoist/winch, refer to Hoist-Supported Personnel Lifting Devices.

Personnel Access Platform Hoist/Winch: A dedicated hoist/winch whose only purpose is to raise and lower a personnel access platform not carrying personnel.

Personnel Licensing: A means to ensure an individual is qualified to perform a designated task.

Proof Load: The specific load or weight applied in performance of a proof load test (typically greater than the rated load of the LDE).

Proof Load Test: A load test performed prior to first use, after major modification of the load path, or at other prescribed times. This test verifies material strength, construction, and workmanship and typically uses a load greater than the rated load.

Qualified Person: A person who, by possession of a recognized degree in an applicable field or certificate of professional standing, or who, by extensive knowledge, training, and experience, has successfully demonstrated the ability to solve or resolve problems relating to the subject matter and work.

Rated Capacity: See Rated Load.

Rated Load: The maximum load a lifting device or equipment is designed to lift under normal operating conditions. This value may be marked on the device indicating maximum capacity. This is also the load referred to as "safe working load or the working load limit." If the device has never been downrated or uprated, this also is the "manufacturer rated load."

Recognized Licensing Organization: Operator licensing organization meeting industry recognized criteria for written testing materials, practical examinations, test administration, grading, facilities/equipment, and personnel.

Regular Service LDE: LDE used one or more times per month.

Remote Emergency Stop (Remote E-Stop): A manually operated switch or valve to cut off electric or fluid power independently of the regular operating controls that is located remotely from the operator control station.

Request for Relief: Documented request for permission to perform some act contrary to established requirements.

Responsible Organization: Entity or a representative thereof responsible for the design, operation, maintenance, testing, inspection, or personnel training and licensing of LDE. (In some cases, this may be the LDEM).

Rigging Hardware: A detachable load supporting device such as a shackle, link, eyebolt, ring, swivel, or clevis.

Safe Working Load: See Rated Load.

Safety Factor: See Design Factor.

Single Failure Point: A single item or component whose failure would cause an undesired event such as dropping a load or loss of control.

Sling: A flexible lifting assembly and incorporated hardware used between a lifting device and the payload being lifted. Common types include wire rope slings, synthetic roundslings, metal mesh slings, synthetic web slings, and chain slings.

Standby LDE: LDE not in regular service but used occasionally or intermittently as required. A lifting device or equipment that has not been used for a period of 1 month or more but less than 12 months is considered to be used intermittently/occasionally.

Structural Sling: A term sometimes used for rigid or semi-rigid lifting devices such as spreader bars or lifting beams that now are included in the general category of below-the-hook lifting devices.

Surface Nondestructive Testing: Test and inspection methods used to examine the surface of equipment/materials (e.g., magnetic particle and liquid penetrant).

Two-Block: A condition in which the lower load block or hook assembly comes into contact with the upper load block, hoist/trolley structure, or boom point sheave assembly.

Volumetric Nondestructive Testing: Test and inspection methods used to examine the interior of equipment/materials (e.g., ultrasonic and radiographic).

Voluntary Consensus Standards (VCS): Industry standards used by NASA for LDE design, operations, maintenance, and inspections, including American Gear Manufacturers Association (AGMA), American Society of Mechanical Engineers (ASME), Deutsches Institut für Normung (DIN), American National Standards Institute (ANSI), and Scaffold and Access Industry Association (SAIA).

Winch: A stationary motor-driven or hand-powered hoisting machine having a drum around which is wound a rope, chain, or web used for lifting and lowering a load (requirements in this standard do not apply to winches used for horizontal pulls).

4. GENERAL LDE REQUIREMENTS

4.1 General

4.1.1   This chapter contains general LDE requirements. Subsequent chapters provide additional requirements specific to individual types of LDE.

4.1.2   The Center/Facility LDE program must be managed in accordance with NPR 8715.3, NASA General Safety Program Requirements, and the requirements in this standard.

> *Note: NPR 8715.3, NASA General Safety Program Requirements, establishes the roles and responsibilities for NASA LDE programs.*

4.1.3   LDE shall be designed, constructed, tested, inspected, maintained, and operated in accordance with the applicable OSHA regulations, the requirements in this standard, VCS as specified herein, and be based upon manufacturer recommendations.

> *Note: Chapter 2 contains a reference list of OSHA regulations and VCS. Refer to the specific LDE chapters for any additional requirements.*

4.1.4   LDEM approval shall be obtained for any tailoring of manufacturer recommendations.

4.2 Classification of Lifts

4.2.1   There are two categories of lifting operations for the purposes of this standard: critical lifts and noncritical lifts. Requirements for critical lifts and critical LDE are specifically addressed throughout the document and are referenced in Appendix B.

4.2.2   The responsible organization shall follow a documented process that seeks input from the appropriate stakeholders (such as facility, program, operations, and safety) and the LDEM to classify lifts as critical or noncritical.

4.2.3   The LDEM shall have the authority to reclassify noncritical lifts as critical based upon safety, facility, or other concerns beyond normal lifting operations.

4.2.4   An operation shall be classified as a critical lift when failure/loss of control presents an elevated risk of serious injury, loss of life, or loss of one-of-a-kind articles, high dollar items or major facility components whose loss would have serious programmatic or institutional impact.

> *Note: Lifts of high-value spacecraft are usually classified as critical lifts, while lifts of small, improvised mini satellites, for example, most likely would not be. Lifting and movement of flight hardware components packaged per applicable shipment specifications are typically not classified as critical lifts.*

4.2.5   An operation may be classified as a noncritical lift if it does not meet critical lift criteria.

Note: Noncritical lifts typically involve routine lifting operations and are governed by standard industry rules and practices except as supplemented with unique NASA testing, operations, maintenance, inspection, and personnel licensing requirements contained in this standard.

4.3   Safety Hazard Analysis

4.3.1   A recognized safety hazard analysis shall be performed on critical or custom-built LDE (subject to documented LDEM approval, hooks, rigging hardware, slings, below-the-hook lifting devices, and attachments for industrial trucks may be excluded for cases in which there is no potential for load instability).

Note: Some one-of-a-kind, custom-built LDE may be more likely to break down and should be considered less reliable than commercial off the shelf (COTS) equipment. Given this, original equipment manufacturer (OEM)-type LDE should be used when possible rather than custom, built-up equipment.

4.3.2   The safety hazard analysis shall, as a minimum, identify potential sources of danger and recommend resolutions for those conditions that could cause loss of life, personal injury, and loss of or damage to the LDE, facility, or load.

4.4   Design

4.4.1   In accordance with paragraph 4.1.3, LDE must be designed and constructed in accordance with the applicable OSHA regulations, the requirements in this standard, and VCS as specified herein.

4.4.2   If critical or custom-built LDE is designed or procured, the responsible organization shall notify the LDEM and provide the LDEM with the necessary information for review and approval of the design/procurement prior to use of the hardware (subject to documented LDEM approval, hooks, rigging hardware, and slings may be excluded from this requirement).

4.5   Testing

4.5.1   As stated in paragraph 4.1.3, tests must comply with the applicable OSHA regulations, the requirements in this standard, the applicable VCS as specified herein, and be based upon manufacturer recommendations. In accordance with paragraph 4.1.4, tailoring of manufacturer recommendations for testing requires LDEM approval.

4.5.2   Two types of tests are specified for LDE in this standard: proof load tests and periodic load tests. The required tests and test parameters will vary according to the specific LDE. Refer to the applicable LDE chapter for specific requirements and additional information.

Note: Periodic load tests are limited to a value below that of proof tests to avoid damage to the rope and overheating of drive components.

4.5.3 Designated persons shall perform all load tests in accordance with written procedures.

4.5.4 LDE shall undergo a proof load test:

a. Prior to first use for all new LDE.

b. Prior to being placed back into service after repairs or modifications that affect load holding capability, such as welding on components in the load path.

c. After wire ropes or load chains are replaced.

4.5.5 When a proof load test is required, a periodic load test shall also be performed.

Note 1: For industrial trucks and their attachments, load measuring devices, and jacks, the proof load test consists of performing a periodic load test.

Note 2: For slings, rigging hardware, and below-the-hook lifting devices, performance of a proof load test satisfies the periodic load test requirement for that load test cycle.

4.5.6 Periodic load testing of extensively repaired or modified LDE may be limited to the functions affected by the repair or modification only if the periodic load test interval has not expired.

4.5.7 Repaired or modified LDE components that do not affect the lifting or holding capability of the LDE shall undergo a functional check prior to the LDE being placed back into service to verify the component repairs or modifications are acceptable.

4.5.8 Load testing should be conducted in an area where minimal damage will occur if the LDE fails.

4.5.9 An inspection of the LDE and its components shall be performed prior to and after each load test to ensure there is no damage before releasing the LDE into service.

4.5.10 Tests shall be current before Idle and Standby LDE are returned to service.

Note: Testing and inspection are not required while LDE is in Idle or Standby status.

4.6 Inspection

4.6.1 In accordance with paragraph 4.1.3, daily, frequent, periodic, and pre-use inspections must be performed in accordance with this standard, applicable OSHA regulations and VCS, and be based upon manufacturer recommendations. In accordance with paragraph 4.1.4, tailoring of manufacturer recommendations for inspections requires LDEM approval.

4.6.2 Designated persons shall conduct all LDE inspections.

4.6.3   Periodic inspections shall be conducted in accordance with written procedures.

4.6.4   A periodic inspection shall be performed on all new, extensively repaired, or extensively modified LDE prior to first use.

> *Note: For component repair on LDE, only the inspections that apply to the repaired portion need to be performed prior to first use if the periodic inspection interval has not expired.*

4.6.5   Inspections shall be current before Idle and Standby LDE is returned to service.

> *Note: Testing and inspections are not required while LDE is in Idle or Standby LDE status.*

4.7   Operation

4.7.1   In accordance with paragraph 4.1.3, LDE operations must comply with this standard, applicable OSHA regulations and VCS, and be based upon manufacturer recommendations. In accordance with paragraph 4.1.4, tailoring of manufacturer recommendations for operations requires LDEM approval.

4.7.2   For all lifts, a designated person responsible for the safety of the operation shall be present.

> *Note: For routine lifts involving minimal risk, the equipment operator may serve as the designated person.*

4.7.3   The effects of weather conditions on lift safety shall be evaluated prior to performing LDE operations.

> *Note: Operations are generally permitted without restriction during electrical storms within enclosed metal or framed buildings that are properly grounded.*

4.7.4   LDE found in an unsafe operating condition shall be removed from service.

4.7.5   LDE problems/discrepancies shall be documented and dispositioned prior to use.

4.7.6   LDE shall be verified to be within inspection and testing intervals prior to use.

4.7.7   LDE shall not be loaded beyond its rated load except for required testing.

> *Note 1: Follow applicable OSHA regulations when testing LDE.*
> *Note 2: 29 CFR 1926 requires knowledge of the weight of the load.*

4.7.8   If radio or other communications are to be used, operators or lift supervisors shall test the communication system prior to each operation. Operations shall stop immediately upon communication loss and shall not continue until communication is restored.

4.7.9   Specific written procedures shall be prepared and followed for critical lifts.

4.7.10   Before lifting flammable or energetic materials such as explosives, pyrotechnics, or propellants, the need for special provisions such as additional bonding, grounding, and stray voltage testing shall be assessed.

> Note: Refer to NASA-STD-8719.12, Safety Standard for Explosives, Propellants, and Pyrotechnics, for requirements applicable to explosives handling and processing.

4.8   Maintenance

4.8.1   In accordance with paragraph 4.1.3, LDE maintenance programs must comply with this standard, applicable OSHA regulations and VCS, and be based upon manufacturer recommendations. In accordance with paragraph 4.1.4, tailoring of manufacturer recommendations for maintenance requires LDEM approval.

4.8.2   The maintenance program shall include procedures and a scheduling system for normal periodic maintenance items, adjustments, replacements, and repairs.

4.8.3   Maintenance safety precautions shall be taken in accordance with OSHA, the applicable VCS, and be based upon manufacturer recommendations.

4.8.4   LDEM approval shall be obtained for any modifications to LDE.

> Note: Replacement in kind is not considered a modification and does not require LDEM approval.

4.9   Labeling and Tagging

4.9.1   In accordance with paragraph 4.1.3, labeling and tagging of LDE must comply with this standard, applicable OSHA regulations, and VCS, and be based upon manufacturer recommendations.

4.9.2   The rated load shall be plainly marked on LDE.

> Note 1: For some types of equipment, a capacity plate affixed to the LDE or a load chart kept on the LDE is acceptable. Consult OSHA and applicable VCS.

> Note 2: Hooks that are part of other LDE, and attachments that are permanently mounted on industrial trucks, do not need separate marking.

4.9.3   Critical LDE shall be marked conspicuously as such.

> Note: Hooks that are part of critical LDE, and attachments that are permanently mounted on industrial trucks, do not need separate marking.

4.9.4 Following each periodic load test, a durable tag shall be affixed to the LDE identifying the equipment and stating the next required periodic load test date or load test expiration date.

Note 1: See Chapter 14 Slings, Rigging Hardware, and Below-the-Hook Lifting Devices for additional requirements.

Note 2: Hooks that are part of other LDE and attachments that are permanently mounted on industrial trucks do not need separate marking.

4.9.5 Idle and Standby LDE shall be conspicuously marked as such.

4.10 Records

4.10.1 The responsible organization shall ensure:

a. Test, inspection, and maintenance records comply with the applicable OSHA regulations and VCS.

b. Records of each test and periodic inspection are generated.

Note: Consult OSHA and VCS for additional documentation requirements.

c. LDE maintenance records are generated.

d. LDE record retention is in accordance with NRRS 1441.1, NASA Records Retention Schedules.

e. LDE and its status are tracked and controlled using a configuration management system.

4.11 Personnel Training and Licensing

4.11.1 General

4.11.1.1 Personnel operating LDE shall be appropriately trained and licensed.

Note: LDE operators must be appropriately trained. This standard does not require LDE operators be licensed to operate manually operated hoists and winches, personnel access platform hoists/winches, manually propelled mobile aerial platforms (e.g., access stand/stairs), manually propelled industrial trucks, manually operated load positioning devices, load measuring devices, and jacks, but additional licensing may be required by Center policy or the LDEM.

4.11.1.2 A training, examination, and licensing program shall be established.

Note: For those NASA installations not having a training program, LDE operators may be trained and licensed by a recognized licensing organization.

4.11.1.3   Licenses shall indicate the type of LDE the holder is qualified and authorized to operate.

Note: The responsible organization may elect to maintain a master list of licensed operators instead of issuing individual licenses, provided copies of the list are readily available to assurance and supervisory personnel at the work site.

4.11.1.4   Rigging shall be performed by designated persons.

Note: LDE operators may perform rigging tasks for which they are trained and qualified.

4.11.1.5   Nondestructive testing (NDT) personnel shall be licensed in accordance with a nationally or internationally recognized NDT personnel qualifications practice or standards such as ASNT-CP-189, SNT-TC-1A, NAS-410, or a similar document.

Note 1: Routine visual inspections that are part of daily, frequent, periodic, and other LDE inspections as outlined in OSHA, VCS, and this document are not considered NDT for the purposes of personnel licensing.

Note 2: Visual acuity requirements are included in all NDT personnel licensing standards.

4.11.1.6   Signal persons shall be trained on the types and application of signals and LDE operations.

4.11.2   Licensing Program

4.11.2.1   Licensing programs must comply with the applicable OSHA regulations and shall be based upon VCS and manufacturer recommendations.

Note: Refer to [Chapter 2, Applicable Documents](#).

4.11.2.2   A responsible organization shall oversee the issuance of personnel licenses.

4.11.2.3   Licensing organizations and the LDEM shall reserve the right to suspend or revoke licenses for reasons such as negligence, violations of requirements, or failure to meet medical standards documented in NPR 1800.1.

4.11.2.4   Initial licensing training and examination for LDE operators shall include the following as a minimum:

a.   Training in safety, lifting equipment emergency procedures, general performance standards, requirements, pre-operational checks, and safety-related defects and symptoms.

b.   Hands-on training.

c.   Written examination.

d.   Operational demonstration.

e. Physical examination of licensed personnel in accordance with NPR 1800.1, NASA Occupational Health Program Procedures.

4.11.2.5   Licenses shall expire every four years or less, contingent upon maintenance of a current physical examination in accordance with NPR 1800.1, NASA Occupational Health Program Procedures.

4.11.2.6   Licensing organizations shall establish renewal procedures that include:

a. A written examination and operational demonstration, at a minimum.

b. Training in safety, lifting equipment emergency procedures, general performance standards, requirements, pre-operational checks, and safety-related defects and symptoms and hands-on training, as needed.

c. Verification of compliance with NPR 1800.1, NASA Occupational Health Program Procedures, requirements regarding physical examination of licensed personnel.

4.11.2.7   Licensing shall be revoked if personnel do not maintain compliance with licensing requirements.

4.11.2.8   The LDEM shall review the personnel licensing program at least annually to ensure the contents, training material, testing, and examination elements are up-to-date with current methods and techniques and any "lessons-learned" are adequately addressed.

5.   OVERHEAD CRANES

5.1   General

5.1.1   The requirements contained in this chapter are applicable to overhead cranes including gantry, top running or underhung, monorail, and jib cranes.

5.1.2   In accordance with paragraph 4.1.3, design, construction, testing, inspection, maintenance, and operation of overhead cranes must comply with the applicable OSHA regulations, the requirements in this standard, ASME B30 series standards (ASME B30.2, ASME B30.11, ASME B30.17, or ASME B30.24) and CMAA Specification 70 or 74 or equivalent as approved by the LDEM. Additionally, operation, testing, inspection, and maintenance of overhead cranes must be based upon manufacturer recommendations.

5.1.3   As stated in section 1.4, OSHA and other regulatory requirements take precedence in case of conflict. The requirements in this document take precedence over VCS except in those cases in which the VCS is invoked by regulation. In case of questions regarding conflicting requirements or to request a clarification, contact the LDEM.

5.2 Classification of Lifts

Classify lifts in accordance with section 4.2. There are no additional requirements specific to overhead cranes in this section.

5.3 Safety Hazard Analysis

Perform a safety hazard analysis on critical or custom-built equipment as required in section 4.3. There are no additional requirements specific to overhead cranes in this section.

5.4 Design

As stated in paragraph 4.1.3, design and construction must comply with the applicable OSHA regulations, the VCS specified at the beginning of this chapter, the requirements in section 4.4, and the following:

5.4.1 Mechanical

    5.4.1.1   Cranes used for critical lifts shall have one of the following:

a. Two holding brakes, each capable of bringing a rated load to zero speed and holding it.

Note: A load brake may be considered a second holding brake provided it is capable of bringing a rated load to zero speed and holding it.

b. A single holding brake in combination with a motor drive that automatically monitors brake functionality and motor torque.

    5.4.1.2   Holding brake(s) shall be applied automatically when power to the brake is removed.

    5.4.1.3   The brake design should provide for emergency load lowering.

    5.4.1.4   When used for critical lifts, speed reduction from the motor to the drum on the hoist should be achieved by using gears enclosed in a gear case. If open gears are required, they shall be guarded, with provision for lubrication and inspection.

    5.4.1.5   Worm gears shall not be used as a braking means unless the lead angle prevents back driving.

Note: The braking properties of a worm gear tend to degrade with use; the design engineer should consider this in existing installations where the hoist is subject to heavy use or when purchasing new equipment.

    5.4.1.6   Cast iron components shall not be used in the hoist load path unless approved by the LDEM and the responsible organization.

Note: The material properties of cast iron allow catastrophic failure (brittle fracture), and it should not be considered as reliable as steel or cast steel. The engineer should consider this when selecting equipment and avoid the use of load bearing cast iron materials where possible.

5.4.1.7 Crane design shall provide for visual and physical accessibility for safe inspection, service, repair, and component replacement.

5.4.2 Electrical

5.4.2.1 Emergency stops (E-Stops) shall open the mainline contactor or the main circuit breaker.

Note: Emergency lighting and other personnel safety circuits may remain powered after remote E-stop actuation.

5.4.2.2 Operator E-stops shall be controlled by a red pushbutton accessible to the operator.

5.4.2.3 In cases where the operator's view is restricted/obstructed, the requirements of paragraph 5.7.6 apply.

5.4.2.4 Remote E-Stops shall be:

a. Located such that the E-Stop operator(s) can clearly see the load and lift area(s).

b. Operated separately from and take precedence over the operator control circuit.

c. Operated by a standardized hand-held remote E-Stop pendant that includes power and circuit continuity indications.

5.4.2.5 Cranes used for critical lifts shall be equipped with dual upper limit switches.

5.4.2.6 For critical lift electric cranes, the limit switches shall meet the following:

a. The initial upper limit switch precludes movement in the raise direction when the limit is reached.

Note: Movement in the "lower" direction need not be inhibited in association with the initial upper limit switch function.

b. The final upper limit switch is wired into the mainline circuit, hoist power circuit, main contactor control circuit, or hoist power contactor control circuit, such that all crane motion or all hoist motion is precluded when the limit is reached.

c. After a final upper limit switch has been activated, movement of the load requires action (resetting) at the final upper limit switch level.

Note: The crane design should include a means of detecting limit switch failure and allow for safe inspection and repair. For example, a system may be equipped with two different colored annunciator lights, one for each limit switch. A reset button may be included so when a final upper limit switch is tripped, the load can be lowered immediately. The reset button should be secured to prevent unauthorized or unintended use.

d. The initial upper limit switch is adjusted sufficiently low to preclude inadvertent actuation of the final upper limit switch if the hoist actuates the initial upper limit switch at full speed with no load. Similarly, the final upper limit switch is adjusted sufficiently low to ensure the hoist will not two-block (or otherwise damage wire rope) if the hoist actuates the final upper limit switch at full speed with no load.

Note: This requirement effectively lowers the usable hook height of the hoist.

5.4.2.7   For cranes used for critical lifts, a lower limit switch shall be provided to ensure no less than two wraps remain on the drum.

Note: Movement in the "raise" direction need not be inhibited in association with the lower limit switch function.

5.4.2.8   Critical lift cranes should have a fail-safe control system such that a single failure does not cause the crane to operate at a speed faster than commanded or in a direction other than commanded.

Note: A failure that stops the crane and sets the brakes or causes the crane to operate at a speed slower than commanded without disabling the stop function is acceptable.

5.5   Testing

As stated in paragraph 4.1.3, tests must comply with the applicable OSHA regulations, the VCS specified at the beginning of this chapter, the requirements in this section and section 4.5, and be based upon manufacturer recommendations.

5.5.1   Proof Load Test

5.5.1.1   Proof load tests, as stipulated in section 4.5, shall be performed with a dummy load of 1.20 to 1.25 times the rated capacity of the crane.

5.5.1.2   Proof load tests shall be conducted after the crane has been installed at the site or facility in which it will be used.

5.5.1.3   Loads shall be held for a time sufficient to verify no drift occurs.

5.5.1.4   Refer to section 13.6 for hook NDT requirements.

5.5.2   Periodic Load Test

In accordance with section 4.5, a periodic test must be performed whenever a proof load test is required.

5.5.2.1   A periodic load test shall be performed on each crane at least once every four years.

5.5.2.2   A periodic load test shall have been performed on a crane within one year prior to its use for a critical lift.

5.5.2.3   Periodic load test intervals may be extended by no more than 90 days from the original expiration date due to programmatic or institutional needs, subject to LDEM approval. To extend the periodic load test interval, the following conditions shall be met:

a. The responsible organization provides documented rationale to the LDEM.

b. The LDEM determines there is no increase in risk.

5.5.2.4   The periodic load test shall consist of the following:

a. With a dummy load equal to 1.00 to 1.05 times the crane's rated capacity:

   (1) Raise and lower the load at various speeds to ensure the hoist is functional under load.

   Note: Consult the LDEM regarding appropriate range of travel.

   (2) Travel the load at various speeds to ensure the bridge and trolley are functional under load.

   Note: Consult the LDEM regarding appropriate range of travel.

b. Test the holding brakes in one of the following ways:

   (1) Statically test each brake (under no load) to the design rated torque at the point of brake application.

   Note: This method is preferred.

   (2) Check each brake for its ability to hold a static dummy load equal to 1.00 to 1.05 times the crane's rated capacity.

   Note 1: It must be possible to reactivate the out-of-circuit brake.

   Note 2: If a worm gear or a load brake is used as a holding brake, test to ensure it is able to hold a static rated load.

   (3) Other methods as approved by the LDEM.

c. Test all E-Stop switches with no load on the hook by operating the E-stop and verifying all crane motions are precluded.

d. Test all limit switches with no load on the hook by operating the crane at slow speed into the limit switch and verifying the appropriate crane motion is precluded.

Note: For cranes equipped with dual upper hoist limit switches, the final upper limit switch may be tested by manually tripping the switch and verifying all hoist motion is precluded.

e. Test safety devices when possible.

Note: It is not always possible to test safety devices (e.g., circuit breakers and thermal overload protection).

5.5.2.5   Refer to section 13.6 for hook NDT requirements.

5.6 Inspection

As stated in paragraph 4.1.3, inspections must comply with the applicable OSHA regulations, the VCS specified at the beginning of this chapter, the requirements in section 4.6, and be based upon manufacturer recommendations. There are no additional requirements specific to overhead cranes in this section.

5.7 Operation

5.7.1   As stated in paragraph 4.1.3, operations must comply with the applicable OSHA regulations, the VCS specified at the beginning of this chapter, the requirements in this section and section 4.7, and be based upon manufacturer recommendations.

5.7.2   Methods and procedures should be developed for lowering a load in the event of crane failure or other contingencies. These should be demonstrated and verified if practical.

5.7.3   Use of overhead cranes for load testing items such as slings, platforms, and lifting fixtures or to relieve a portion of the weight of a constrained load shall be subject to the following conditions:

a. The crane is specifically identified and documented for such use and approved by the LDEM.

b. A load measuring device is installed in the lifting assembly.

c. The crane is not used to load test items by pulling against the ground or against an otherwise fixed object.

d. The total measured load on the crane is not to exceed 50 percent of the rated capacity of the crane when performing the following activities:

(1) Load testing an item at a load greater than its periodic load test value as stipulated in this document.

Note: Refer to paragraph 14.5.2.2 for additional restrictions on testing certain types of slings above their rated capacities.

(2) Load testing an item by pulling against an object whose weight exceeds the desired test load.

(3) Relieving a portion of the weight of a constrained load.

e. The load is only applied vertically.

f. When load testing an item freely suspended from the hook, the test weight is not to be lifted more than six inches above the floor/working surface or above the lowest reasonable height based on test item dimensions and configuration, subject to approval by the LDEM.

g. The crane or lifting assembly (e.g., load positioning device) is to have sufficient fine motion capability to precisely control movement of the load so as to avoid crane overload or damage to the item when performing the following activities:

(1) Load testing an item by pulling against an object whose weight exceeds the desired test load.

(2) Relieving a portion or all of the weight of a constrained load.

h. When a crane is used to load test an item by pulling against an object whose weight exceeds the desired test load, the weight of the object is to be within the rated capacity of the crane.

5.7.4 OSHA requires testing the brakes when raising loads that approach the rated capacity of the crane.

5.7.5 If conventional means of reaching a worksite such as an aerial platform, ladder, stairs, or scaffold would be more hazardous, or if access is not possible because of structural design or worksite conditions, and it is determined personnel must be lifted with a crane, the requirements of 29 CFR 1926 for mobile cranes shall be followed.

5.7.6 One of the following options shall be implemented for lifts where the operator's view is restricted/obstructed:

a. One or more remote E-Stops as required to ensure safe operations (see E-Stop requirements in paragraph 5.4.2).

b. Handling procedures that minimize the risk, with LDEM approval.

Note: Remote E-Stops are the preferred method.

5.7.7 Any time a final upper limit switch is activated, the cause shall be determined and resolved prior to further operations.

5.7.8 Personnel shall not be located under a suspended load except as specifically authorized by the OSHA-approved NASA Alternate Standard for Suspended Load Operations (see Appendix A).

> *Note: In accordance with Appendix A, a list of approved suspended load operations, a list of cranes/hoists used for suspended load operations, and copies of the associated hazards analyses must be provided to the OSHA Office of Federal Agency Programs via NASA Headquarters for distribution to the appropriate regional and area OSHA offices. Quarterly updates to the documentation will be provided as needed.*

5.8 Maintenance

As stated in paragraph 4.1.3, maintenance must comply with the applicable OSHA regulations, the VCS specified at the beginning of this chapter, the requirements in section 4.8, and be based upon manufacturer recommendations. There are no additional requirements specific to overhead cranes in this section.

5.9 Labeling and Tagging

5.9.1 As stated in paragraph 4.1.3, labeling and tagging must comply with the applicable OSHA regulations, the VCS specified at the beginning of this chapter, the requirements in section 4.9, and be based upon manufacturer recommendations.

5.9.2 Overhead cranes shall have the directions of bridge and trolley travel marked on the cranes such that:

    a. The directions correspond to the directions on the operator controls.

    b. The markings are visible from ground level.

> *Note: Directional markings may be omitted when the crane height or other conditions exist such that reading the markings is difficult or impractical, subject to LDEM approval.*

5.10 Records

As stated in paragraph 4.1.3, record generation and retention must comply with the applicable OSHA regulations, the VCS specified at the beginning of this chapter, the requirements in section 4.10, and be based upon manufacturer recommendations. There are no additional requirements specific to overhead cranes in this section.

5.11 Personnel Training and Licensing

In accordance with section 4.11, personnel training and licensing must comply with the applicable OSHA regulations, the requirements in this standard, and be based upon VCS and manufacturer recommendations. There are no additional requirements specific to overhead cranes in this section.

6. MOBILE CRANES AND DERRICKS

6.1 General

6.1.1 The requirements contained in this chapter are applicable to mobile cranes and derricks.

6.1.2 In accordance with paragraph 4.1.3, design, construction, testing, inspections, maintenance, and operations of mobile cranes and derricks must comply with the applicable OSHA regulations, the requirements in this standard, and the applicable ASME B30 series or DIN standards (ASME B30.5, ASME B30.6, DIN EN 13000) or equivalent as approved by the LDEM. Additionally, operation, testing, inspection, and maintenance of mobile cranes must be based upon manufacturer recommendations.

6.1.3 As stated in section 1.4, OSHA and other regulatory requirements take precedence in case of conflict. The requirements in this document take precedence over VCS except in those cases in which the VCS is invoked by regulation. In case of questions regarding conflicting requirements or to request a clarification, contact the LDEM.

6.2 Classification of Lifts

Classify lifts in accordance with section 4.2. There are no additional requirements specific to mobile cranes and derricks in this section.

6.3 Safety Hazard Analysis

Perform a safety hazard analysis on critical or custom-built equipment as required in section 4.3. There are no additional requirements specific to mobile cranes and derricks in this section.

6.4 Design

As stated in paragraph 4.1.3, design and construction must comply with the applicable OSHA regulations, the VCS specified at the beginning of this chapter, and the requirements in section 4.4. There are no additional requirements specific to mobile cranes and derricks in this section.

6.5 **Testing**

As stated in paragraph 4.1.3, tests must comply with the applicable OSHA regulations, the VCS specified at the beginning of this chapter, the requirements in this section and section 4.5, and be based upon manufacturer recommendations.

6.5.1   Proof Load Test

6.5.1.1   Proof load tests as stipulated in section 4.5 shall be performed with a dummy load of 0.95 to 1.00 times the rated capacity of the mobile crane/derrick at the maximum and minimum working radius, except as provided in paragraph 6.5.1.2.

> *Note 1: Proof load tests for mobile cranes conducted by the manufacturer prior to delivery are acceptable if load test documentation is provided to verify the extent and thoroughness of the test.*
>
> *Note 2: Testing at the minimum practical working radius that does not interfere with the crane/derrick structure is acceptable.*

6.5.1.2   The dummy load may exceed 1.00 times the rated capacity of the mobile crane/derrick, subject to approval. To exercise this option, the following conditions shall be met:

   a. The responsible organization obtains written approval from the manufacturer or a qualified person.

   b. The responsible organization provides documented rationale to the LDEM.

   c. The LDEM provides approval.

6.5.1.3   Loads shall be held for a time sufficient to verify no drift occurs.

6.5.1.4   Refer to section 13.6 for hook NDT requirements.

6.5.1   Periodic Load Test

6.5.2.1   In accordance with section 4.5, a periodic load test must be performed whenever a proof load test is required.

6.5.2.2   A periodic load test shall be performed on each mobile crane/derrick at least once every four years.

6.5.2.3   A periodic load test shall have been performed on a mobile crane/derrick within one year prior to its use for a critical lift.

6.5.2.4   Periodic load test intervals may be extended by no more than 90 days from the original expiration date due to programmatic or institutional needs, subject to LDEM approval. To extend the periodic load test interval, the following conditions shall be met:

a. The responsible organization provides documented rationale to the LDEM.

b. The LDEM determines there is no increase in risk.

6.5.2.5 A periodic load test shall be performed after each boom change (when boom disassembly/assembly is required) if the mobile crane/derrick is to be used for critical lifts.

6.5.2.6 The periodic load test shall consist of the following:

a. With a dummy load equal to 0.95 to 1.00 times the rated capacity at the minimum practical working radius:

(1) Hoist and lower the load at various speeds with the boom at the minimum radius.
> *Note: Testing at the minimum practical working radius that does not interfere with the crane/derrick structure is acceptable.*

(2) Hold the load for a sufficient duration to verify no drift occurs.
> *Note: The load should be held long enough to allow any dynamics to dampen out.*

b. Check hoist brake system functionality by placing the load on the hook, hoisting up a few inches, holding the load for a time sufficient for the power-controlled lowering mechanism to bleed off fluid (verifies the functionality of the holding brake), then slowly lowering the load to the ground (verifies proper operation of the power-controlled lowering mechanism).

c. With a dummy load not less than 0.50 times the rated capacity at a radius that will safely clear the outriggers (for telescopic boom cranes, use a boom length where all sections are partially extended, if possible):

(1) Perform boom hoisting and lowering.

(2) Check swing mechanism operation, pausing at each outrigger (when so equipped) for sufficient duration to verify no drift occurs.

d. With no load on the hook:

(1) Test all limit switches and E-stop switches.

(2) Test locking devices, boom angle indicators, and other safety devices when possible.
> *Note: It is not always possible to test safety devices (e.g., circuit breakers and thermal overload protection).*

6.5.2.7   Other methods may be used to satisfy the periodic load test requirements, as approved by the LDEM.

6.5.2.8   Refer to section 13.6 for hook NDT requirements.

6.6   Inspection

As stated in paragraph 4.1.3, inspections must comply with the applicable OSHA regulations, the VCS specified at the beginning of this chapter, the requirements in section 4.6, and be based upon manufacturer recommendations. There are no additional requirements specific to mobile cranes and derricks in this section.

6.7   Operation

6.7.1   As stated in paragraph 4.1.3, operations must comply with the applicable OSHA regulations, the VCS specified at the beginning of this chapter, the requirements in this section and section 4.7, and be based upon manufacturer recommendations.

6.7.2   The operator shall verify the weight of the load to be within the rated capacity of the crane/derrick by at least one of the following methods:

> (1) From a source recognized by the industry (such as the load's manufacturer), or from a calculation method recognized by the industry (such as calculating a steel beam from measured dimensions and a known per foot weight), or from other equally reliable means.
>
> *Note: The operator may request to view this information prior to the lift and may refuse to perform the lift if this information is not provided.*
>
> (2) By beginning to lift the load and using a load weighing device, load moment indicator, rated capacity indicator, or rated capacity limiter to determine the following:
>
>> a. If the load weight is below 75 percent of the maximum rated capacity at the longest radius that will be used during the operation the operator may proceed with the lift.
>>
>> b. If the load weight exceeds 75 percent of the maximum rated capacity at the longest radius that will be used during the operation, the operator shall not proceed with the lift until the weight of the load is verified in accordance with 6.7.2(1).

6.7.3   Use of mobile cranes and derricks for load testing items such as slings, platforms, and lifting fixtures or to relieve a portion of the weight of a constrained load shall be subject to the following conditions:

> a. The crane/derrick is specifically identified and documented for such use and approved by the LDEM.

b. The load is not to exceed 75 percent of the rated capacity of the crane.

c. The load is only applied vertically.

d. A load-measuring device is installed in the lifting assembly.

e. The boom angle is minimized as much as safely possible to prevent the boom from contacting the boom stops.

f. The boom shall be adequately restrained to prevent damage to the crane due to sudden unloading should the test article fail.

   Note: Hydraulic boom cranes with sufficient damping to adequately restrain the boom (such as when equipped with double acting cylinders) do not need additional means of boom restraint, subject to approval by the LDEM.

g. When load testing an item freely suspended from the hook, the test weight is not to be lifted more than six inches above the floor/working surface or above the lowest reasonable height based on test item dimensions and configuration, subject to approval by the LDEM.

h. When load testing an item by pulling against a constrained object or one whose weight exceeds the desired test load, or when using a mobile crane/derrick to relieve a portion or all of the weight of a constrained load, the crane/derrick or lifting assembly (e.g., load positioning device) shall have sufficient fine motion capability to control movement of the load precisely so as to avoid crane overload or damage to the load.

   Note: Using a mobile crane to load test items by pulling against an object whose weight exceeds the desired test load or for relieving a portion or all of the weight of a constrained load should be avoided if possible.

6.7.4 OSHA requires testing the brakes when raising loads that approach the rated capacity of the crane.

6.7.5 If conventional means of reaching a worksite such as an aerial platform, ladder, stairs, or scaffold would be more hazardous, or if access is not possible because of structural design or worksite conditions, and it is determined that personnel must be lifted with a crane, the requirements of 29 CFR 1926 shall be followed for lifting of personnel with a crane.

6.7.6 Personnel shall not be located under a suspended load except as specifically authorized by the OSHA-approved NASA Alternate Standard for Suspended Load Operations (see Appendix A).

   Note: In accordance with Appendix A, a list of approved suspended load operations, a list of cranes/hoists used for suspended load operations, and copies of the associated hazards analyses must be provided to the OSHA Office of Federal Agency Programs via NASA

Headquarters for distribution to the appropriate regional and area OSHA offices. Quarterly updates to the documentation will be provided as needed.

6.7.7　ASME B30.5 states that verified weights, measured radii, and manufacturer load rating chart capacities and instructions take precedence over operational aids.

6.7.8　ASME B30.5 states the load chart must be used to confirm the crane configuration is adequate for the load, the site, and lift conditions.

6.8　Maintenance

As stated in paragraph 4.1.3, maintenance must comply with the applicable OSHA regulations, the VCS specified at the beginning of this chapter, the requirements in section 4.8, and be based upon manufacturer recommendations. There are no additional requirements specific to mobile cranes and derricks in this section.

6.9　Labeling and Tagging

As stated in paragraph 4.1.3, labeling and tagging must comply with the applicable OSHA regulations, the VCS specified at the beginning of this chapter, the requirements in section 4.9, and be based upon manufacturer recommendations. There are no additional requirements specific to mobile cranes and derricks in this section.

6.10　Records

As stated in paragraph 4.1.3, record generation and retention must comply with the applicable OSHA regulations, the VCS specified at the beginning of this chapter, the requirements in section 4.10, and be based upon manufacturer recommendations. There are no additional requirements specific to mobile cranes and derricks in this section.

6.11　Personnel Training and Licensing

In accordance with section 4.11, personnel training and licensing must comply with the applicable OSHA regulations, the requirements in this standard, and be based upon VCS and manufacturer recommendations. There are no additional requirements specific to mobile cranes and derricks in this section.

7.　HOISTS AND WINCHES

7.1　General

7.1.1　The requirements contained in this chapter are applicable to hoists and winches used for lifting and lowering a load but do not apply to winches used for horizontal pulls. These requirements apply to electric, air-powered, and manual hoists and winches including personnel access platform hoists/winches whose only purpose is to raise and lower a platform not carrying personnel. Additional requirements for hoists connected to platforms used to raise or lower personnel are contained in Chapter 8, Hoist-Supported Personnel Lifting Devices.

7.1.2 There are no OSHA regulations pertaining specifically to hoists and winches for general industry. However, in accordance with paragraph 4.1.3, design, construction, testing, inspection, maintenance, and operation of hoists and winches must comply with the applicable OSHA regulations, the requirements in this standard, and ASME B30 series standards (ASME B30.7, ASME B30.16, or ASME B30.21) or equivalent as approved by the LDEM. Additionally, operation, testing, inspection, and maintenance of hoists and winches must be based upon manufacturer recommendations.

7.1.3 As stated in section 1.4, OSHA and other regulatory requirements take precedence in case of conflict. The requirements in this document take precedence over VCS except in those cases in which the VCS is invoked by regulation. In case of questions regarding conflicting requirements or to request a clarification, contact the LDEM.

7.2 Classification of Lifts

Classify lifts in accordance with section 4.2. There are no additional requirements specific to hoists and winches in this section.

7.3 Safety Hazard Analysis

Perform a safety hazard analysis on critical or custom-built equipment as required in section 4.3. There are no additional requirements specific to hoists and winches in this section.

7.4 Design

As stated in paragraph 4.1.3, design and construction must comply with the applicable OSHA regulations, the VCS specified at the beginning of this chapter, the requirements in section 4.4, and the following:

7.4.1 Mechanical

7.4.1.1 For powered hoists and winches:

   a. Except as provided in paragraph 7.4.1.1.b, hoists/winches used for critical lifts shall have one of the following:

   (1) Two holding brakes, each capable of bringing a rated load to zero speed and holding it.

   Note: A load brake may be considered a second holding brake provided it is capable of bringing a rated load to zero speed and holding it.

(2) A single holding brake in combination with a motor drive that automatically monitors brake functionality and motor torque.

b. Subject to LDEM approval, a hoist/winch with a single holding brake may be used for critical lifts when hoists/winches compliant with paragraphs 7.4.1.1.a(1) and (2) are not commercially available. To exercise this option, the following conditions shall be met:

(1) The responsible organization provides documented rationale to the LDEM.

Note: Rationale may include design and construction information, inspections, operations, maintenance and storage provisions, or other considerations.

(2) The LDEM determines there is no increase in risk.

c. Holding brake(s) shall be applied automatically when power to the brake is removed.

d. The brake design should provide for emergency load lowering.

e. When used for critical lifts, speed reduction from the motor to the drum on the hoist should be achieved by using gears enclosed in a gear case. If open gears are required, they shall be guarded with a provision for lubrication and inspection.

7.4.1.2 Worm gears shall not be used as a holding brake unless the lead angle prevents back driving.

Note: The braking properties of a worm gear tend to degrade with use; the design engineer should consider this in existing installations where the hoist is subject to heavy use or when purchasing new equipment.

7.4.1.3 Cast iron components shall not be used in the hoist or winch load path unless approved by the LDEM and the responsible organization.

Note: The material properties of cast iron allow catastrophic failure (brittle fracture), and it should not be considered as reliable as steel or cast steel. The engineer should consider this when selecting equipment and avoid the use of load bearing cast iron materials where possible.

7.4.1.4 Hoist/winch design shall provide for visual and physical accessibility for safe inspection, service, repair, and component replacement.

7.4.2 Over-Travel Protection for Powered Hoist and Winches

7.4.2.1 Except as provided in paragraph 7.4.2.5 and 7.4.2.6, electric and air-powered hoists and winches used for critical lifts shall be equipped with dual upper limit switches.

7.4.2.2   Except as provided in paragraph 7.4.2.5, for electric and air-powered hoists and winches used for critical lifts, a lower limit switch shall be provided to ensure no less than two wraps remain on the drum.

> *Note: Movement in the "raise" direction need not be inhibited in association with the lower limit switch function.*

7.4.2.3   For air-powered hoists and winches equipped with dual upper limit switches, the final upper limit switch shall exhaust air from the hoist or winch, set the brakes, and require reset at the upper limit switch level.

7.4.2.4   Electric hoists and winches equipped with dual upper limit switches shall meet the following:

a. The initial upper limit switch precludes movement in the raise direction when the limit is reached.

> *Note: Movement in the "lower" direction need not be inhibited in association with the initial upper limit switch function.*

b. The final upper limit switch is wired into the mainline circuit, hoist/winch power circuit, main contactor control circuit, or hoist/winch power contactor control circuit such that all hoist/winch motion is precluded when the limit is reached.

c. After a final upper limit switch has been activated, movement of the load requires action (resetting) at the final upper limit switch level.

> *Note: The hoist design should include a means of detecting limit switch failure and allow for safe inspection and repair. For example, a system may be equipped with two different colored annunciator lights, one for each limit switch. A reset button may be included so that when a final upper limit switch is tripped, the load can be lowered immediately. The reset button should be secured to prevent unauthorized or unintended use.*

d. The initial upper limit switch is adjusted sufficiently low to preclude inadvertent actuation of the final upper limit switch if the hoist actuates the initial switch at full speed with no load. Similarly, the final upper limit is adjusted sufficiently low to ensure the hoist or winch will not two-block (or otherwise damage wire rope) if the hoist or winch actuates the final switch at full speed with no load.

> *Note: This requirement effectively lowers the usable hook height of the hoist.*

7.4.2.5 Subject to LDEM approval, a hoist/winch with a single upper limit and no lower limit switch may be used for critical lifts when hoists/winches compliant with paragraphs 7.4.2.1 and 7.4.2.2 are not commercially available. To exercise this option, the following conditions shall be met:

a. The responsible organization provides documented rationale to the LDEM.

   Note: Rationale may include design and construction information, inspections, operations, maintenance and storage provisions, or other considerations.

b. The LDEM determines there is no increase in risk.

7.4.2.6 Air-powered chain hoists may use a travel-limiting clutch in place of the final upper limit switch.

7.4.3   E-Stops for Powered Hoists and Winches

7.4.3.1   For electric hoists and winches, E-Stops shall open the mainline contactor or the main circuit breaker.

Note: Emergency lighting and other personnel safety circuits may remain powered after remote E-stop actuation.

7.4.3.2   For air-powered hoists and winches, E-stops shall remove or isolate the pneumatic source from the hoist/winch.

Note: A dump valve is acceptable for the E-Stop provided it also isolates the pneumatic source from the hoist/winch.

7.4.3.3   Operator E-stops shall be controlled by a red pushbutton accessible to the operator.

7.4.3.4   In cases where the operator's view is restricted/obstructed, the requirements of paragraph 7.7.3.b apply.

7.4.3.5   Remote E-Stops shall be:

a. Located such that the E-Stop operator(s) can clearly see the load and lift area(s).

b. Operated separately from and take precedence over the operator control circuit.

c. Operated by a standardized hand-held remote E-Stop pendant that includes power and circuit continuity indications.

7.5   Testing

As stated in paragraph 4.1.3, tests must comply with the applicable OSHA regulations, the VCS specified at the beginning of this chapter, the requirements in this section and section 4.5, and be based upon manufacturer recommendations.

7.5.1 Proof Load Test

7.5.1.1   Proof load tests as stipulated in section 4.5 shall be performed with a dummy load of 1.20 to 1.25 times the rated capacity of the hoist/winch.

Note: For personnel access platform hoists/winches, the attached platform may serve as part of the dummy load.

7.5.1.2   Proof load tests for powered hoists and winches shall be conducted after installation at the site or facility in which they will be used, except as permitted in paragraph 7.5.1.3.c.

7.5.1.3   Proof load tests at the site or facility are not required when replacing a powered hoist/winch on an existing mounting structure when the following conditions are met:

   a. The new hoist/winch is a replacement in kind (manufacturer, model, and load rating).

   b. The existing mounting structure has been previously proof load tested in a manner that meets this standard.

   c. The new hoist has been proof load tested by the manufacturer in a manner that meets this standard.

   d. The responsible organization obtains LDEM approval.

7.5.1.4   Loads shall be held for a time sufficient to verify no drift occurs.

7.5.1.5   Refer to section 13.6 for hook NDT requirements.

7.5.2 Periodic Load Test

7.5.2.1   In accordance with section 4.5, a periodic load test must be performed whenever a proof load test is required.

7.5.2.2   A periodic load test shall be performed on each hoist/winch at least once every four years.

7.5.2.3   A periodic load test shall have been performed on a hoist/winch within one year prior to its use for a critical lift.

7.5.2.4   Periodic load test intervals may be extended by no more than 90 days from the original expiration date due to programmatic or institutional needs, subject to LDEM approval. To extend the periodic load test interval, the following conditions shall be met:

   a. The responsible organization provides documented rationale to the LDEM.

   b. The LDEM determines there is no increase in risk.

7.5.2.5 The periodic load test shall consist of the following:

    a. With a dummy load equal to 1.00 to 1.05 times the hoist/winch's rated capacity or with the attached personnel access platform serving as the dummy load (personnel access platform hoists/winches only), raise and lower the load at various speeds to ensure the hoist is functional under load.

        Note: Consult the LDEM regarding appropriate range of travel.

    b. Test the holding brakes in one of the following ways:

        (1) Statically test each brake (under no load) to the design rated torque at the point of brake application.

            Note: This method is preferred.

        (2) Check each brake for its ability to hold a static dummy load equal to 1.00 to 1.05 times the hoist's/winch's rated capacity.

            Note 1: It must be possible to reactivate the out-of-circuit brake.

            Note 2: If a worm gear or a load brake is used as a holding brake, test to ensure it is able to hold a static rated load.

            Note 3: For personnel access platform hoists/winches, the attached personnel access platform may serve as the dummy load.

        (3) Other methods as approved by the LDEM.

    c. Test E-Stop switches with no load on the hook (for personnel access platform hoists/winches, the personnel access platform may remain attached) by operating the E-stop and verifying all motions are precluded.

    d. Test all limit switches with no load on the hook by operating the hoist/winch at slow speed into the limit switch and verifying the appropriate motion is precluded.

        Note: For hoists/winches equipped with dual upper hoist limit switches, the final upper limit switch may be tested by manually tripping the switch and verifying all hoisting motion is precluded.

    e. Test safety devices when possible.

        Note: It is not always possible to test safety devices (e.g., circuit breakers and thermal overload protection).

7.5.2.6 Refer to section 13.6 for hook NDT requirements.

7.6 Inspection

7.6.1 As stated in paragraph 4.1.3, inspections must comply with the applicable OSHA regulations, the VCS specified at the beginning of this chapter, the requirements in this section and section 4.6, and be based upon manufacturer recommendations.

7.6.2 Each day a manual lever-operated hoist is used, the following daily inspections shall be performed:

   a. Check operating mechanisms for proper operation, proper adjustments, and unusual sounds.

   b. Check load-bearing components for damage (including hooks, chain, rope, or web straps).

7.7 Operation

7.7.1 As stated in paragraph 4.1.3, operations must comply with the applicable OSHA regulations, the VCS specified at the beginning of this chapter, the requirements in this section and section 4.7, and be based upon manufacturer recommendations.

7.7.2 Methods and procedures should be developed for lowering a load in the event of hoist failure or other contingencies. These should be demonstrated and verified if practical.

7.7.3 One of the following options shall be implemented for lifts using powered hoists when the operator's view is restricted/obstructed:

   a. One or more remote E-Stops as required to ensure safe operations (see E-Stop requirements in paragraph 7.4.3).

   b. Handling procedures that minimize the risk, with LDEM approval.

   Note: Remote E-Stops are the preferred method.

7.7.4 Any time a final upper limit switch is activated, the cause shall be determined and resolved prior to further operations.

7.7.5 Hoists and winches shall not be used for lifting personnel unless specifically designed for such a purpose (see Chapter 8).

7.7.6 Before each lift or series of lifts for powered hoists/winches (except for personnel access platform hoists/winches), the operator shall functionally test proper operation of the upper limit switch with no load on the hook.

7.7.7 Upper limit switches shall not be used as operating controls.

7.7.8 Use of hoist/winches for load testing items such as slings, platforms, and lifting fixtures or to relieve a portion of the weight of a constrained load shall be subject to the following conditions:

    a. Hoist/winch is specifically identified and documented for such use and approved by the LDEM.

    b. Hoist/winch is not used to load test items by pulling against the ground or against an otherwise fixed object.

    c. Except in cases where the hoist/winch is used exclusively for load testing and related activities, the total load is not to exceed 50 percent of the rated capacity of the hoist/winch when performing the following:

        (1) Load testing an item at a load greater than its periodic load test value as stipulated in this document.

        Note: Refer to paragraph 14.5.2.2 for additional restrictions on testing certain types of slings above their rated capacity.

        (2) Load testing an item by pulling against an object whose weight exceeds the desired test load.

    d. The total load is not to exceed 50 percent of the rated capacity of the hoist/winch when relieving a portion of the weight of a constrained load.

    e. The load is only applied vertically.

    f. A load measuring device is installed in the lifting assembly.

    g. When load testing an item freely suspended from the hook, the test weight is not to be lifted more than six inches above the floor/working surface or above the lowest reasonable height based on test item dimensions and configuration, subject to approval by the LDEM.

    h. The hoist/winch or lifting assembly (e.g., load positioning device) is to have sufficient fine motion capability to precisely control movement of the load so as to avoid hoist/winch overload or damage to the item when performing the following:

        (1) Load testing an item by pulling against an object whose weight exceeds the desired test load.

        (2) Relieving a portion or all of the weight of a constrained load.

i. When load testing an item by pulling against an object whose weight exceeds the desired test load, the weight of the object is within the rated capacity of the hoist/winch.

7.7.9　ASME B30.7 and B30.16 require testing the brakes when raising loads that approach the rated capacity of the hoist/winch.

7.7.10　Personnel shall not be located under a suspended load except as specifically authorized by the OSHA-approved NASA Alternate Standard for Suspended Load Operations (see Appendix A).

> *Note: In accordance with Appendix A, a list of approved suspended load operations, a list of cranes/hoists used for suspended load operations, and copies of the associated hazards analyses will be provided to the OSHA Office of Federal Agency Programs via NASA Headquarters for distribution to the appropriate regional and area OSHA offices. Quarterly updates to the documentation will be provided as needed.*

7.8　Maintenance

As stated in paragraph 4.1.3, maintenance must comply with the applicable OSHA regulations, the VCS specified at the beginning of this chapter, the requirements in section 4.8, and be based upon manufacturer recommendations. There are no additional requirements specific to hoists and winches in this section.

7.9　Labeling and Tagging

As stated in paragraph 4.1.3, labeling and tagging must comply with the applicable OSHA regulations, the VCS specified at the beginning of this chapter, the requirements in section 4.9, and be based upon manufacturer recommendations. There are no additional requirements specific to hoists and winches in this section.

7.10　Records

As stated in paragraph 4.1.3, record generation and retention must comply with the applicable OSHA regulations, the VCS specified at the beginning of this chapter, the requirements in section 4.10, and be based upon manufacturer recommendations. There are no additional requirements specific to hoists and winches in this section.

7.11　Personnel Training and Licensing

In accordance with section 4.11, personnel training and licensing must comply with the applicable OSHA regulations, the requirements in this standard, and be based upon VCS and manufacturer recommendations. Per paragraph 4.11.1.1, operators of manually operated hoists/winches and personnel access platform hoists must be appropriately trained but do not have to be licensed unless required by Center policy or the LDEM. There are no additional requirements specific to hoists and winches in this section.

8. HOIST-SUPPORTED PERSONNEL LIFTING DEVICES

8.1 General

8.1.1 The requirements contained in this chapter are applicable to hoist-supported personnel lifting devices. This chapter applies to devices specifically designed to lift and lower persons via hoist, including hoist-supported platforms where personnel occupy the platform during movement. This chapter does not apply to the following:

   a. Personnel access platforms hoisted unoccupied to a position and anchored or restrained to a stationary structure before personnel occupy the platform.

   b. Elevators covered by ASME A17.1, "Elevators, Dumbwaiters, Escalators, and Moving Walks."

   c. Mobile aerial platforms (e.g., manlifts, aerial devices, scissors lifts, or other devices covered by ANSI/SAIA A92 series standards). See Chapter 9 for Mobile Aerial Platforms.

   d. Platforms covered by ASME A120.1, "Safety Requirements for Powered Platforms and Traveling Ladders and Gantries for Building Maintenance."

8.1.2 There are no OSHA regulations or VCS specifically addressing hoist-supported personnel lifting devices covered by this chapter. However, in accordance with paragraph 4.1.3, hoist-supported personnel lifting devices must meet the applicable OSHA regulations and VCS for design, construction, testing, inspection, maintenance, and operation that apply to the LDE of which they are composed, in addition to the requirements in this standard. Additionally, operation, testing, inspection, and maintenance of hoist-supported personnel lifting devices must be based upon manufacturer recommendations for the LDE of which they are composed.

8.2 Classification of Lifts

8.2.1 Classify lifts in accordance with section 4.2. There are no additional requirements specific to hoist-supported personnel lifting devices in this section.

8.2.2 Lifting of personnel using a hoist-supported personnel lifting device shall be classified as a critical lift.

8.3 Safety Hazard Analysis

8.3.1 Perform a safety hazard analysis on critical or custom-built equipment as required in section 4.3. There are no additional requirements specific to hoist-supported personnel lifting devices in this section.

8.3.2 A recognized safety hazard analysis shall be performed on hoist-supported personnel lifting devices.

8.4 Design

8.4.1 As stated in paragraph 4.1.3, hoist-supported personnel lifting devices must comply with the applicable OSHA regulations and VCS for the LDE of which they are composed, the requirements of this standard, including those in section 4.4, section 7.4, and the following:

8.4.2 Hoist-supported personnel lifting devices shall have at least one of the following:

   a. Two independent support systems consisting of two separate hoists such that the failure of one hoist, its reeving system, or other component will not cause the stability of the personnel lifting device to be lost or prohibit its movement to a safe location.

   b. A single support system with two or more holding brakes and additional design factors for the hoist and other load bearing components as approved by the LDEM.

   c. Other methods/attributes as approved by the LDEM.

8.4.3 Hoist-supported personnel lifting devices shall allow for safe egress of personnel being lifted or for emergency lowering to the ground level or other safe location.

8.4.4 Hoist-supported personnel lifting devices shall be equipped with an E-Stop device within reach of the person controlling movement of the device that deenergizes the powered systems and stops the movement of the device.

   Note: An additional E-Stop separate from normal operating controls should be considered for personnel at ground level or on a fixed structure to enhance operational safety.

8.4.5 All directional controls shall be designed so they automatically return to a neutral position when released. Neutral position of controls shall bring the unit to a safe stop and hold the unit in that position until commanded to move to another position.

8.5 Testing

As stated in paragraph 4.1.3, tests of hoist-supported personnel lifting devices must comply with the applicable OSHA regulations and VCS for the LDE of which they are composed, the requirements in this section and section 4.5, and be based upon manufacturer recommendations.

8.5.1 Proof Load Test

8.5.1.1 Proof load tests as stipulated in section 4.5 shall be performed with a dummy load of 1.45 to 1.50 times the rated capacity of the hoist-supported personnel lifting devices.

8.5.1.2 Loads shall be held for a time sufficient to verify no drift occurs.

8.5.1.3   Refer to section 13.6 for hook NDT requirements.

8.5.2   Periodic Load Test

8.5.2.1   In accordance with section 4.5, a periodic load test must be performed whenever a proof load test is required.

8.5.2.2   A periodic load test shall be performed on each hoist-supported personnel lifting device at least once every year with a load equal to 1.00 to 1.05 times the device's rated load.

8.5.2.3   Periodic load test intervals may be extended by no more than 90 days from the original expiration date due to programmatic or institutional needs, subject to LDEM approval. To extend the periodic load test interval, the following conditions shall be met:

   a. The responsible organization provides documented rationale to the LDEM.

   b. The LDEM determines there is no increase in risk.

8.5.2.4   The periodic load test shall consist of the following:

   a. Raise and lower the load at various speeds with a dummy load equal to 1.00 to 1.05 times the personnel lifting device's rated capacity to ensure the hoist(s) is functional under load.

   Note: Consult the LDEM regarding appropriate range of travel.

   b. Test the holding brakes in one of the following ways:

      (1) Statically test each brake (under no load) to the design rated torque at the point of brake application.

      Note: This method is preferred.

      (2) Check each brake for its ability to hold a static dummy load equal to 1.00 to 1.05 times the rated capacity of the personnel lifting device.

      Note 1: It must be possible to reactivate the out-of-circuit brake.

      Note 2: If a worm gear or a load brake is used as a holding brake, test to ensure it is able to hold a static rated load.

      (3) Other methods as approved by the LDEM.

   c. Test E-Stop switches with no load on the personnel lifting device by operating the E-stop and verifying all motions are precluded.

d. Test all limit switches with no load on the device by operating the device at slow speed into the limit switch and verifying the appropriate motion is precluded.

> *Note: The final upper limit switch may be tested by manually tripping the switch and verifying all hoist motion is precluded.*

e. Test safety devices when possible.

> *Note: It is not always possible to test safety devices (e.g., circuit breakers and thermal overload protection).*

8.5.2.5 Refer to section 13.6 for hook NDT requirements.

8.6 Inspection

As stated in paragraph 4.1.3, inspections of hoist-supported personnel lifting devices must comply with the applicable OSHA regulations and VCS for the LDE of which they are composed, the requirements in this section and section 4.6, and be based upon manufacturer recommendations.

8.6.1 Daily Inspection

Each day a hoist-supported personnel lifting device is used, the following inspections shall be performed prior to using the device:

a. Check for defects such as cracked welds, damaged control cables, loose wire connections, and wheel or roller damage.

b. Check operating mechanisms, control mechanisms, and guard rails for proper function.

c. Check hose and fittings, tanks, valves, drain pumps, gear casings, and other components of fluid systems for deterioration and leaks.

d. Without disassembling, inspect all functional operating and control mechanisms for excessive wear and contamination by excessive lubricants or foreign matter.

8.6.2 Periodic Inspection

At least once per year, the following inspections shall be performed on all hoist-supported personnel lifting devices, in addition to a daily inspection:

a. Check for deformed, cracked, or corroded members and welds and loose bolts or rivets in the personnel lift structure.

b. Check for cracked or worn sheaves and drums.

c. Check for excessive wear or cracks in pins, bearings, shafts, gears, followers, and locking and clamping devices.

8.7 Operation

8.7.1 As stated in paragraph 4.1.3, operation of hoist-supported personnel lifting devices must comply with the applicable OSHA regulations and VCS for the LDE of which they are composed, the requirements in this section and section 4.7, and be based upon manufacturer recommendations.

8.7.2 The operator shall perform a pre-operational check of the controls. If controls do not operate properly, repairs and adjustments shall be made before operations begin.

8.7.3 Materials and equipment shall be secured while the platform is lifted.

8.7.4 Prior to an operation, hoist-supported personnel lifting device operators shall test the communication system. If communications are disrupted, all operations shall be stopped until communication is reestablished.

8.7.5 Personnel shall keep all parts of the body, tools, and equipment inside the platform periphery during raising, lowering, and traveling operations.

8.8 Maintenance

As stated in paragraph 4.1.3, maintenance of hoist-supported personnel lifting devices must comply with the applicable OSHA regulations and VCS for the LDE of which they are composed, the requirements in section 4.8, and be based upon manufacturer recommendations. There are no additional requirements specific to hoist-supported personnel lifting devices in this section.

8.9 Labeling and Tagging

As stated in paragraph 4.1.3, labeling and tagging of hoist-supported personnel lifting devices must comply with the applicable OSHA regulations and VCS for the LDE of which they are composed, the requirements in section 4.9, and be based upon manufacturer recommendations. There are no additional requirements specific to hoist-supported personnel lifting devices in this section.

8.10 Records

As stated in paragraph 4.1.3, record generation and retention of hoist-supported personnel lifting devices must comply with the applicable OSHA regulations and VCS for the LDE of which they are composed, the requirements in section 4.10, and be based upon manufacturer recommendations. There are no additional requirements specific to hoist-supported personnel lifting devices in this section.

8.11 Personnel Training and Licensing

In accordance with section 4.11, personnel training and licensing of hoist-supported personnel lifting devices must comply with the applicable OSHA regulations and be based upon VCS and manufacturer recommendations for the LDE of which they are composed. There are no additional requirements specific to hoist-supported personnel lifting devices in this section.

9. MOBILE AERIAL PLATFORMS

9.1 General

9.1.1 The requirements contained in this chapter are applicable to mobile aerial platforms, including vehicle mounted elevating and rotating aerial devices, manually propelled elevating aerial platforms, boom supported elevating work platforms, and self-propelled elevating work platforms.

9.1.2 In accordance with paragraph 4.1.3, design, construction, testing, inspections, maintenance, and operations of mobile aerial platforms must comply with the applicable OSHA regulations, the requirements in this standard, and ANSI/SAIA A92 series standards (ANSI/SAIA A92.2, A92.3, A92.5, A92.6) or equivalent as approved by the LDEM. Additionally, operation, testing, inspection, and maintenance of mobile aerial platforms must be based upon manufacturer recommendations.

9.1.3 As stated in section 1.4, OSHA and other regulatory requirements take precedence in case of conflict. The requirements in this document take precedence over VCS except in those cases in which the VCS is invoked by regulation. In case of questions regarding conflicting requirements or to request a clarification, contact the LDEM.

9.2 Classification of Lifts

Classify lifts in accordance with section 4.2. There are no additional requirements specific to mobile aerial platforms in this section.

9.3 Safety Hazard Analysis

Perform a safety hazard analysis on critical or custom-built equipment as required in section 4.3. There are no additional requirements specific to mobile aerial platforms in this section.

9.4 Design

As stated in paragraph 4.1.3, design and construction must comply with the applicable OSHA regulations, the VCS specified at the beginning of this chapter, and the requirements in section 4.4. There are no additional requirements specific to mobile aerial platforms in this section.

9.5 Testing

As stated in paragraph 4.1.3, tests must comply with the applicable OSHA regulations, the VCS specified at the beginning of this chapter, the requirements in this section and section 4.5, and be based upon manufacturer recommendations.

9.5.1 Proof Load Test

Proof load tests as stipulated in section 4.5 shall be performed in accordance with manufacturer instructions and the applicable VCS.

9.5.2 Periodic Load Test

9.5.2.1 In accordance with section 4.5, a periodic load test must be performed whenever a proof load test is required.

9.5.2.2 A periodic load test shall be performed on each mobile aerial platform at least once a year.

9.5.2.3 Periodic load test intervals may be extended by no more than 90 days from the original expiration date due to programmatic or institutional needs, subject to LDEM approval. To extend the periodic load test interval, the following conditions shall be met:

   a. The responsible organization provides documented rationale to the LDEM.

   b. The LDEM determines there is no increase in risk.

9.5.2.4 The periodic load test shall consist of the following:

   a. Hold a dummy load equal to 1.00 to 1.05 times the device's rated capacity (at maximum boom radius, when applicable) for a sufficient duration to verify drift does not exceed that specified by the responsible organization.

      Note: Equipment application, manufacturer recommendations, and engineering analyses should be taken into account when determining acceptable amount of drift.

   b. Test all functions in an unloaded condition, including operation of limit switches and tilt alarm/shutoff.

      Note: Where possible, use ground control station. When it is necessary to use the platform control station, operate close to ground level.

9.6 Inspection

As stated in paragraph 4.1.3, inspections must comply with the applicable OSHA regulations, the VCS specified at the beginning of this chapter, the requirements in section 4.6, and be based upon manufacturer recommendations. There are no additional requirements specific to mobile aerial platforms in this section.

9.7 Operation

As stated in paragraph 4.1.3, operations must comply with the applicable OSHA regulations, the VCS specified at the beginning of this chapter, the requirements in section 4.7, and be based upon manufacturer recommendations. There are no additional requirements specific to mobile aerial platforms in this section.

9.8 Maintenance

As stated in paragraph 4.1.3, maintenance must comply with the applicable OSHA regulations, the VCS specified at the beginning of this chapter, the requirements in section 4.8, and be based upon manufacturer recommendations. There are no additional requirements specific to mobile aerial platforms in this section.

9.9 Labeling and Tagging

As stated in paragraph 4.1.3, labeling and tagging must comply with the applicable OSHA regulations, the VCS specified at the beginning of this chapter, the requirements in section 4.9, and be based upon manufacturer recommendations. There are no additional requirements specific to mobile aerial platforms in this section.

9.10 Records

As stated in paragraph 4.1.3, record generation and retention must comply with the applicable OSHA regulations, the VCS specified at the beginning of this chapter, the requirements in section 4.10, and be based upon manufacturer recommendations. There are no additional requirements specific to mobile aerial platforms in this section.

9.11 Personnel Training and Licensing

In accordance with section 4.11, personnel training and licensing must comply with the applicable OSHA regulations, the requirements in this standard, and be based upon VCS and manufacturer recommendations. Per paragraph 4.11.1.1, manually propelled mobile aerial platform operators must be appropriately trained, but do not have to be licensed unless required by Center policy or the LDEM. There are no additional requirements specific to mobile aerial platforms in this section.

10. HIGH LIFT INDUSTRIAL TRUCKS

10.1 General

10.1.1 The requirements contained in this chapter are applicable to high lift industrial trucks including forklift trucks, platform trucks, picker trucks, and reach trucks. This chapter is not applicable to low lift trucks (such as pallet jacks).

10.1.2 In accordance with paragraph 4.1.3, design, construction, testing, inspections, maintenance, and operations of high lift industrial trucks and their attachments must comply with the applicable OSHA regulations, the requirements in this standard, and the applicable ANSI/ITSDF standards (ANSI/ITSDF B56.1, B56.6, B56.10, or B56.14) or equivalent as approved by the LDEM. Additionally, operation, testing, inspection, and maintenance of high lift industrial trucks and their attachments must be based upon manufacturer recommendations.

10.1.3 As stated in section 1.4, OSHA and other regulatory requirements take precedence in case of conflict. The requirements in this document take precedence over VCS except in those cases in which the VCS is invoked by regulation. In case of questions regarding conflicting requirements or to request a clarification, contact the LDEM.

10.2 Classification of Lifts

Classify lifts in accordance with section 4.2. There are no additional requirements specific to high lift industrial trucks in this section.

10.3 Safety Hazard Analysis

10.3.1 Perform a safety hazard analysis on critical or custom-built equipment as required in section 4.3, except with documented authorization by the LDEM in accordance with paragraph 4.3.1.

10.3.2 There are no additional requirements specific to high lift industrial trucks and their attachments in this section.

10.4 Design

As stated in paragraph 4.1.3, design and construction of equipment must comply with the applicable OSHA regulations, the VCS specified at the beginning of this chapter, and the requirements in section 4.4. There are no additional requirements specific to high lift industrial trucks in this section.

10.5 Testing

As stated in paragraph 4.1.3, tests must comply with the applicable OSHA regulations, the VCS specified at the beginning of this chapter, the requirements in this section and section 4.5, and be based upon manufacturer recommendations.

10.5.1 Proof Load Test

Proof load tests as stipulated in section 4.5 must be performed. The proof load test for industrial trucks and their attachments shall consist of a periodic load test as specified in paragraph 10.5.2.

10.5.2 Periodic Load Test

10.5.2.1 In accordance with section 4.5, performance of a periodic load test satisfies the proof load test requirement for industrial trucks and their attachments.

10.5.2.2 A periodic load test shall have been performed on an industrial truck and its attachments within one year prior to their use for a critical lift.

10.5.2.3 Periodic load test intervals may be extended by no more than 90 days from the original expiration date due to programmatic or institutional needs, subject to LDEM approval. To extend the periodic load test interval, the following conditions shall be met:

  a. The responsible organization provides documented rationale to the LDEM.

  b. The LDEM determines there is no increase in risk.

10.5.2.4 For industrial trucks, the periodic load test shall be performed with a dummy load equal to 1.00 to 1.05 times the industrial truck's rated capacity as follows:

  a. Perform all functions, including tilt operation. Ensure the load is secured and will not move during tilting operations.

  b. Hold the load for a sufficient duration to verify drift does not exceed that specified by the responsible organization.

      Note: Equipment application, manufacturer recommendations, and engineering analyses should be taken into account when determining acceptable amount of drift.

10.5.2.5 For attachments, the periodic load test shall be performed with a dummy load equal to 1.00 to 1.05 times the lesser of the following:

    a. The rated capacity of the attachment.

    Note: If load testing an attachment to its rated capacity, ensure that the industrial truck or test fixture used for load testing can safely handle the load. The capacity of the attachment may exceed the capacity of the industrial truck and attachment combination. Always follow the industrial truck nameplate.

    b. The rated capacity of the industrial truck and attachment combination.

    Note: For attachments used on multiple industrial trucks, test the industrial truck and attachment combination with the highest rated capacity.

10.5.2.6 For attachments, the load shall be held for a minimum of 30 seconds or for a duration as determined by the manufacturer.

10.6 Inspection

10.6.1 As stated in paragraph 4.1.3, inspections must comply with the applicable OSHA regulations, the VCS specified at the beginning of this chapter, the requirements in this section and section 4.6, and be based upon manufacturer recommendations.

10.6.2 At least once per year, the following items shall be visually inspected for excessive deterioration, wear, signs of malfunctions, or other potential problems or discrepancies that may affect the safe operation of the industrial truck:

    a. Frame members.

    b. Welds.

    c. Axle stops.

    d. Safe operating features or devices designed and approved for hazardous area operations.

    e. Motors.

    f. Hydraulic system.

    g. Electrical equipment.

    h. Attachments.

i. Lift and tilt mechanisms.

j. Chains and cables.

k. Tires.

l. Brakes.

m. Steering mechanism.

n. Lights.

o. Warning and safety devices.

p. Controls.

10.7 Operation

As stated in paragraph 4.1.3, operations must comply with the applicable OSHA regulations, the VCS specified at the beginning of this chapter, the requirements in section 4.7, and be based upon manufacturer recommendations. There are no additional requirements specific to high lift industrial trucks in this section.

10.8 Maintenance

As stated in paragraph 4.1.3, maintenance must comply with the applicable OSHA regulations, the VCS specified at the beginning of this chapter, the requirements in section 4.8, and be based upon manufacturer recommendations. There are no additional requirements specific to high lift industrial trucks in this section.

10.9 Labeling and Tagging

As stated in paragraph 4.1.3, labeling and tagging must comply with the applicable OSHA regulations, the VCS specified at the beginning of this chapter, the requirements in section 4.9, and be based upon manufacturer recommendations. OSHA and the applicable VCS require that industrial trucks be marked to identify approved attachments. There are no additional requirements specific to high lift industrial trucks in this section.

10.10 Records

As stated in paragraph 4.1.3, record generation and retention must comply with the applicable OSHA regulations, the VCS specified at the beginning of this chapter, the requirements in section 4.10, and be based upon manufacturer recommendations. There are no additional requirements specific to high lift industrial trucks in this section.

10.11 Personnel Training and Licensing

10.11.1   In accordance with section 4.11, personnel training and licensing must comply with the applicable OSHA regulations, the requirements in this standard, and be based upon VCS and manufacturer recommendations.

10.11.2   Per paragraph 4.11.1.1, manually propelled industrial truck operators must be appropriately trained but do not have to be licensed unless required by Center policy or the LDEM. There are no additional requirements in this section.

10.11.3   In addition to paragraph 4.11.2.1, a performance evaluation of powered industrial truck operators must be performed in accordance with OSHA 1910.178.

11. LOAD POSITIONING AND LOAD MEASURING DEVICES

11.1 General

11.1.1 This chapter contains requirements for load positioning and load measuring devices. Load positioning devices (e.g., Hydra Sets®) are self-contained links between the hoist and the load which provide accurate vertical positioning capability.

11.1.2 There are no OSHA regulations specifically addressing load positioning and load measuring devices. However, in accordance with paragraph 4.1.3, design, construction, testing, inspections, maintenance, and operations of load positioning and load measuring devices must comply with the requirements in this standard, and ASME B30 series standards (ASME B30.20 for load positioning devices, and B30.26 for load measuring devices) or equivalent as approved by the LDEM. Additionally, operation, testing, inspection, and maintenance of load positioning and load measuring devices must be based upon manufacturer recommendations.

11.1.3 As stated in section 1.4, OSHA and other regulatory requirements take precedence in case of conflict. The requirements in this document take precedence over VCS except in those cases in which the VCS is invoked by regulation. In case of questions regarding conflicting requirements or to request a clarification, contact the LDEM.

11.2 Classification of Lifts

Classify lifts in accordance with section 4.2. There are no additional requirements specific to load positioning and load measuring devices in this section.

11.3 Safety Hazard Analysis

Perform a safety hazard analysis on critical or custom-built equipment as required in section 4.3. There are no additional requirements specific to load positioning and load measuring devices in this section.

11.4 Design

11.4.1 As stated in paragraph 4.1.3, design and construction must comply with the VCS specified at the beginning of this chapter, the requirements in section 4.4, and the following:

11.4.2 Pneumatically-controlled load positioning devices shall have the following:

  a. A fail-safe check valve on the pneumatic feed line that "locks up" the device in the event of a drop or loss of pneumatic control system pressure.

  b. A fast acting safety shutoff valve downstream of the load regulator to provide positive control of the device when no motion is desired.

11.5 Testing

As stated in paragraph 4.1.3, tests must comply with the VCS specified at the beginning of this chapter, the requirements in this section and section 4.5, and be based upon manufacturer recommendations.

11.5.1 Proof Load Test

11.5.1.1 Proof load tests as stipulated in section 4.5 must be performed.

11.5.1.2 For load positioning devices, the proof load test shall consist of holding a dummy load of 1.20 to 1.25 times the rated capacity or as recommended by the designer with concurrence from the LDEM.

11.5.1.3 For load measuring devices, the proof load test shall consist of a periodic load test as specified in paragraph 11.5.2.7.

11.5.2 Periodic Load Test

11.5.2.1 In accordance with section 4.5, load positioning devices require a periodic load test whenever a proof load test is performed. For load measuring devices, performance of a periodic load test satisfies the proof load test requirement.

11.5.2.2 A periodic load test shall be performed on each load positioning and load measuring device at least once every four years.

11.5.2.3 A periodic load test shall have been performed on load positioning and load measuring devices within one year prior to their use for a critical lift.

11.5.2.4 Periodic load test intervals may be extended by no more than 90 days from the original expiration date due to programmatic or institutional needs, subject to LDEM approval. To extend the periodic load test interval, the following conditions shall be met:

a. The responsible organization provides documented rationale to the LDEM.

b. The LDEM determines there is no increase in risk.

11.5.2.5 When load positioning device seals are replaced, a load test shall be performed.

11.5.2.6 For load positioning devices, the periodic load test shall consist of the following:

a. With a dummy load equal to 1.00 to 1.05 times the device's rated capacity, operate the unit to the midtravel position. Hold the load for a sufficient duration to verify drift does not exceed that specified by the responsible organization.

   Note: Equipment application, manufacturer recommendations, and engineering analyses should be taken into account when determining acceptable amount of drift.

b. Inspect unit for hydraulic leaks.

c. Inspect for structural damage, scoring, and corrosion of the piston rod.

11.5.2.7 For load measuring devices, the periodic load test shall consist of calibration of the unit per ASME B30.26.

11.6 Inspection

As stated in paragraph 4.1.3, inspections must comply with the VCS specified at the beginning of this chapter, the requirements in this section and section 4.6, and be based upon manufacturer recommendations.

11.6.1 Daily Inspection

11.6.1.1 Prior to use of a load positioning device each day, the following inspections shall be performed:

a. Check operating and control mechanisms for proper function.

b. Without disassembly, visually inspect all functional operating and control mechanisms for excessive wear and contamination by excessive lubricants or foreign matter.

c. Visually inspect for corrosion, damage, cracks, and deformities.

d. Inspect hydraulic system for deterioration and leakage.

e. Check for loose hardware.

11.6.2 Frequent and Periodic Inspections

For load positioning devices, the inspections described in section 11.6.1 shall form part of the frequent and periodic inspections in addition to the inspections in the referenced VCS.

11.7 Operation

11.7.1 As stated in paragraph 4.1.3, operations must comply with the VCS specified at the beginning of this chapter, the requirements in this section and section 4.7, and be based upon manufacturer recommendations.

11.7.2 For pneumatically-controlled load positioning devices, a procedure shall be developed and implemented to ensure the fail-safe check valve is set to an appropriate sensitivity.

> *Note: Normally, the valve is set at the mid-point of its range, which is satisfactory for most operations. However, depending on the specifics of the lift, it may be necessary to reset the valve using a dummy load as outlined in the manufacturer recommended procedures.*

11.8 Maintenance

As stated in paragraph 4.1.3, maintenance must comply with the VCS specified at the beginning of this chapter, the requirements in section 4.8, and be based upon manufacturer recommendations. There are no additional requirements specific to load positioning and load measuring devices in this section.

11.9 Labeling and Tagging

As stated in paragraph 4.1.3, labeling and tagging must comply with the VCS specified at the beginning of this chapter, the requirements in section 4.9, and be based upon manufacturer recommendations. There are no additional requirements specific to load positioning and load measuring devices in this section.

11.10 Records

As stated in paragraph 4.1.3, record generation and retention must comply with the VCS specified at the beginning of this chapter, the requirements in section 4.10, and be based upon manufacturer recommendations. There are no additional requirements specific to load positioning and load measuring devices in this section.

11.11 Personnel Training and Licensing

In accordance with section 4.11, personnel training and licensing must comply with the requirements in this standard and be based upon VCS and manufacturer recommendations. Per paragraph 4.11.1.1, manually operated load positioning device and load measuring device operators must be appropriately trained but do not have to be licensed unless required by Center policy or the LDEM. There are no additional requirements specific to load positioning and load measuring devices in this section.

12. JACKS

12.1 General

12.1.1 The requirements contained in this chapter apply to jacks used for critical lifts and to jacks used for non-critical lifts at the discretion of the LDEM.

> *Note 1: Jacks used for non-critical lifts must comply with OSHA and other regulatory requirements and should be operated, tested, and maintained in accordance with VCS and manufacturer instructions.*

> *Note 2: Jacks used as portions of fixtures whose sole purpose is to render ineffective items such as wheels or casters through minimal lifting are not considered to be lifting hardware and are not subject to the requirements of this chapter.*

12.1.2 In accordance with paragraph 4.1.3, design, construction, testing, inspection, maintenance, and operation of critical lift jacks must comply with the applicable OSHA regulations, the requirements in this standard, and ASME B30 series standards (ASME B30.1) or equivalent as approved by the LDEM. Additionally, operation, testing, inspection, and maintenance of critical lift jacks must be based upon manufacturer recommendations.

12.1.3 As stated in section 1.4, OSHA and other regulatory requirements take precedence in case of conflict. The requirements in this document take precedence over VCS except in those cases in which the VCS is invoked by regulation. In case of questions regarding conflicting requirements or to request a clarification, contact the LDEM.

12.2 Classification of Lifts

12.2.1 Classify lifts in accordance with section 4.2.

12.2.2 When an aircraft lift using jacks meets all of the following criteria, it may be classified as a non-critical lift:

   a. Jacking is performed on an aircraft in accordance with the technical manual provided by the aircraft manufacturer for the specific aircraft.

   b. The technical manual unambiguously identifies the specific jack to be used and provides detailed procedures for its use.

12.3 Safety Hazard Analysis

Perform a safety hazard analysis on critical or custom-built equipment as required in section 4.3. There are no additional requirements specific to jacks in this section.

12.4 Design

As stated in paragraph 4.1.3, design and construction of equipment must comply with the applicable OSHA regulations, the VCS specified at the beginning of this chapter, and the requirements in section 4.4. There are no additional requirements specific to jacks in this section.

12.5 Testing

As stated in paragraph 4.1.3, tests must comply with the applicable OSHA regulations, the VCS specified at the beginning of this chapter, the requirements in this section and section 4.5, and be based upon manufacturer recommendations. There are no additional requirements specific to jacks in this section.

12.5.1 Proof Load Test

Proof load tests as stipulated in section 4.5 must be performed. The proof load test shall consist of a periodic load test as specified in paragraph 12.5.2.4.

> *Note: Proof load tests conducted by the manufacturer prior to delivery are acceptable if load test documentation is provided to verify the extent and thoroughness of the test.*

12.5.2 Periodic Load Test

12.5.2.1 In accordance with section 4.5, performance of a periodic load test satisfies the proof load test requirement for jacks.

12.5.2.2 A periodic load test shall have been performed on each jack within one year prior to its use.

12.5.2.3 Periodic load test intervals may be extended by no more than 90 days from the original expiration date due to programmatic or institutional needs, subject to LDEM approval. To extend the periodic load test interval, the following conditions shall be met:

   a. The responsible organization provides documented rationale to the LDEM.

   b. The LDEM determines there is no increase in risk.

12.5.2.4 The periodic load test shall be performed using a dummy load equal to 1.00 to 1.05 times the jack's rated capacity as follows:

   a. For hydraulic jacks equipped with a mechanical locking ring or equivalent device:

   (1) Check the hydraulic holding capability of the jack by:

       (a) Holding the load using hydraulic pressure at one or more points in the travel, as determined by the responsible organization.

(b) Verifying any drift does not exceed that specified by the responsible organization.

> *Note: Equipment application, manufacturer recommendations, and engineering analyses should be taken into account when determining acceptable amount of drift.*

(2) Check the mechanical locking ring (or equivalent) holding capability of the jack by:

(a) Holding the load using the mechanical locking ring (or equivalent) at one or more points in the travel, as determined by the responsible organization.

(b) Verifying no drift occurs.

b. For hydraulic jacks not equipped with a mechanical locking ring or equivalent device:

(1) Operate to a length of travel as specified by the responsible organization.

(2) Hold the load at one or more points in the travel as determined by the responsible organization.

(3) Verify any drift does not exceed that specified by the responsible organization.

> *Note: Equipment application, manufacturer recommendations, and engineering analyses should be taken into account when determining acceptable amount of drift.*

c. For mechanical jacks:

(1) Operate to a length of travel as specified by the responsible organization.

(2) Hold the load at one or more points in the travel, as determined by the responsible organization.

(3) Verify no drift occurs.

12.6 Inspection

As stated in paragraph 4.1.3, inspections must comply with the applicable OSHA regulations, the VCS specified at the beginning of this chapter, the requirements in section 4.6, and be based upon manufacturer recommendations. There are no additional requirements specific to jacks in this section.

12.7 Operation

As stated in paragraph 4.1.3, operations must comply with the applicable OSHA regulations, the VCS specified at the beginning of this chapter, the requirements in section 4.7, and be based upon manufacturer recommendations. There are no additional requirements specific to jacks in this section.

12.8 Maintenance

As stated in paragraph 4.1.3, maintenance must comply with the applicable OSHA regulations, the VCS specified at the beginning of this chapter, the requirements in section 4.8, and be based upon manufacturer recommendations. There are no additional requirements specific to jacks in this section.

12.9 Labeling and Tagging

As stated in paragraph 4.1.3, labeling and tagging must comply with the applicable OSHA regulations, the VCS specified at the beginning of this chapter, the requirements in section 4.9, and be based upon manufacturer recommendations. There are no additional requirements specific to jacks in this section.

12.10 Records

As stated in paragraph 4.1.3, record generation and retention must comply with the applicable OSHA regulations, the VCS specified at the beginning of this chapter, the requirements in section 4.10, and be based upon manufacturer recommendations. There are no additional requirements specific to jacks in this section.

12.11 Personnel Training and Licensing

In accordance with section 4.11, personnel training and licensing must comply with the applicable OSHA regulations, the requirements in this standard, and be based upon VCS and manufacturer recommendations. Per paragraph 4.11.1.1, jack operators must be appropriately trained, but do not have to be licensed unless required by Center policy or the LDEM. There are no additional requirements specific to jacks in this section.

13. HOOKS

13.1 General

13.1.1 The requirements contained in this chapter are applicable to hooks used on LDE.

13.1.2 In accordance with paragraph 4.1.3, design, construction, testing, inspections, maintenance, and operations using hooks must comply with the applicable OSHA regulations, the requirements in this standard, and ASME B30 series standards (ASME B30.10) or equivalent as approved by the LDEM. Additionally, operation, testing, inspection, and maintenance of hooks must be based upon manufacturer recommendations.

> *Note: Although there are no OSHA regulations pertaining specifically to hooks, they are addressed in various standards that cover the associated equipment or activities in which they are used, such as: 29 CFR 1910.179, 1910.180, 1910.181, 1910.184, 1926 Subpart N, Subpart CC, and others.*

13.1.3 As stated in section 1.4, OSHA and other regulatory requirements take precedence in case of conflict. The requirements in this document take precedence over VCS except in those cases in which the VCS is invoked by regulation. In case of questions regarding conflicting requirements or to request a clarification, contact the LDEM.

13.2 Classification of Lifts

Classify lifts in accordance with section 4.2. There are no additional requirements specific to hooks in this section.

13.3 Safety Hazard Analysis

In accordance with section 4.3, a safety hazard analysis is not required for hooks, subject to LDEM approval. There are no additional requirements specific to hooks in this section.

13.4 Design

As stated in paragraph 4.1.3, design and construction of equipment must comply with the applicable OSHA regulations, the VCS specified at the beginning of this chapter, and the requirements in section 4.4. There are no additional requirements specific to hooks in this section.

13.5 Testing

13.5.1 As stated in paragraph 4.1.3, tests must comply with the applicable OSHA regulations, the VCS specified at the beginning of this chapter, the requirements in this section and section 4.5, and be based upon manufacturer recommendations.

13.5.2 Hooks shall be required to pass the tests of the equipment of which they are a part.

13.5.3 Proof load testing of duplex (sister) hook prongs shall be performed with the applied load at the maximum allowable included angle for the hook design, or as specified by a qualified person.

13.6 Inspection

13.6.1 As stated in paragraph 4.1.3, inspections must comply with the applicable OSHA regulations, the VCS specified at the beginning of this chapter, the requirements in this section and section 4.6, and be based upon manufacturer recommendations.

13.6.2 Hooks shall be inspected during the inspections of the equipment of which they are a part.

13.6.3 Hooks on overhead cranes, mobile cranes, and critical lift hoists shall be inspected using surface NDT immediately after all proof load and periodic load tests prior to further use of the hook.

13.6.4 Periodic hook surface NDT intervals may be extended by no more than four years from the original expiration date due to programmatic or institutional needs, subject to LDEM approval. To extend the surface NDT interval, the following conditions shall be met:

    a. The responsible organization provides documented rationale to the LDEM.

    b. The LDEM determines there is no increase in risk.

13.6.5 Volumetric NDT shall be conducted on new hooks at the discretion of the LDEM and the responsible organization.

13.7 Operation

As stated in paragraph 4.1.3, operations must comply with the applicable OSHA regulations, the VCS specified at the beginning of this chapter, the requirements in section 4.7, and be based upon manufacturer recommendations. There are no additional requirements specific to hooks in this section.

13.8 Maintenance

13.8.1 As stated in paragraph 4.1.3, maintenance must comply with the applicable OSHA regulations, the VCS specified at the beginning of this chapter, the requirements in this section and section 4.8, and be based upon manufacturer recommendations.

13.8.2 After being repaired, hooks shall be proof load tested using the associated lifting device/equipment proof load value.

> *Note: Minor grinding of cracks does not require a proof load test if repairs are made following an approved procedure.*

13.9 Labeling and Tagging

As stated in paragraph 4.1.3, labeling and tagging must comply with the applicable OSHA regulations, the VCS specified at the beginning of this chapter, the requirements in section 4.9, and be based upon manufacturer recommendations. In accordance with section 4.9, hooks that are part of other LDE do not need separate marking, labeling, and tagging. There are no additional requirements specific to hooks in this section.

13.10 Records

As stated in paragraph 4.1.3, record generation and retention must comply with the applicable OSHA regulations, the VCS specified at the beginning of this chapter, the requirements in section 4.10, and be based upon manufacturer recommendations. There are no additional requirements specific to hooks in this section.

13.11 Personnel Training and Licensing

In accordance with section 4.11, personnel training and licensing must comply with the applicable OSHA regulations, the requirements in this standard, and be based upon VCS and manufacturer recommendations. There are no additional requirements specific to hooks in this section.

14. SLINGS, RIGGING HARDWARE, AND BELOW-THE-HOOK LIFTING DEVICES

14.1 General

14.1.1 The requirements contained in this chapter are applicable to slings, rigging hardware, and below-the-hook lifting devices, including slings constructed of wire rope, alloy steel chain, metal mesh, synthetic rope, synthetic web, synthetic round slings, and associated rigging hardware such as shackles, D-rings, turnbuckles, and eyebolts.

14.1.2 In accordance with paragraph 4.1.3, design, construction, testing, inspections, maintenance, and operations of slings must comply with the applicable OSHA regulations, the requirements in this standard, and ASME B30 series standards (ASME B30.9) or equivalent as approved by the LDEM. Additionally, operation, testing, inspection, and maintenance of slings must be based upon manufacturer recommendations.

14.1.3 There are no OSHA regulations specifically addressing rigging hardware or below-the-hook lifting devices. However, in accordance with paragraph 4.1.3, design, construction, testing, inspections, maintenance, and operations of these items must comply with the requirements in this standard, and ASME B30 series standards (B30.20 for below-the-hook lifting devices, and B30.26 for rigging hardware) or equivalent as approved by the LDEM. Additionally, operation, testing, inspection, and maintenance of rigging hardware and below-the-hook lifting devices must be based upon manufacturer recommendations.

14.1.4 As stated in section 1.4, OSHA and other regulatory requirements take precedence in case of conflict. The requirements in this document take precedence over VCS except in those cases in which the VCS is invoked by regulation. In case of questions regarding conflicting requirements or to request a clarification, contact the LDEM.

14.2 Classification of Lifts

Classify lifts in accordance with section 4.2. There are no additional requirements specific to slings, rigging hardware, and below-the-hook lifting devices in this section.

14.3 Safety Hazard Analysis

In accordance with section 4.3, a safety hazard analysis is not required for slings, rigging hardware, and below-the-hook lifting devices, subject to LDEM approval. There are no additional requirements specific to slings, rigging hardware, and below-the-hook lifting devices in this section.

14.4 Design

As stated in paragraph 4.1.3, design and construction must comply with the applicable OSHA regulations, the VCS specified at the beginning of this chapter, the requirements in section 4.4, and the following:

14.4.1 All surfaces of below-the-hook lifting devices not painted, lubricated, or coated with strippable vinyl should be corrosion-protected.

14.4.2 Only accepted industry standard materials and techniques shall be used in slings and rigging hardware (natural rope, knots, and wire rope clips are not considered industry standard materials and techniques).

14.5 Testing

14.5.1 General

14.5.1.1 As stated in paragraph 4.1.3, tests must comply with the applicable OSHA regulations, the VCS specified at the beginning of this chapter, the requirements in this section and section 4.5, and be based upon manufacturer recommendations.

14.5.1.2 When slings and below-the-hook lifting devices are composed of more than one sling or rigging hardware component:

   a. The components shall be tested as an assembly, individually, or both, as dictated by worst case stress and stability considerations.

b. When testing as an assembly, the load test value shall be based upon the rated load for the assembly.

c. When testing as individual components:

(1) Rigging hardware periodic load test intervals may be in accordance with the rigging hardware requirements of this section.

(2) Individual sling and rigging hardware component load test values may be based upon the component rated load within the assembly rather than the individual component rated load.

14.5.1.3 Turnbuckles should be load tested at the open position as a minimum.

Note: It is recommended turnbuckles be tested at the open, closed, and midway positions.

14.5.1.4 For all load tests, the load shall be held for a minimum of 30 seconds or for a duration as determined by the manufacturer.

14.5.1.5 Refer to section 13.6 for hook NDT requirements.

14.5.2 Proof Load Test

14.5.2.1 Proof load tests as stipulated in section 4.5 shall be performed with a dummy load of 2.00 to 2.05 times the rated capacity of slings and rigging hardware, except as specified in 14.5.2.3 and 14.5.2.4.

Note: Batch or production lot testing of sample slings and rigging hardware does not meet the proof load test requirements in this section.

14.5.2.2 Per 29 CFR 1910.184, proof load tests for alloy steel chain, wire rope, metal mesh, synthetic rope, synthetic round slings, and synthetic web slings must only be performed by the manufacturer, the sling's user with written permission and procedures from the manufacturer, or an equivalent entity.

Note: OSHA regulations in 29 CFR 1910.184, Slings, and formal OSHA Letters of Interpretation prohibit loading by the user beyond rated load for alloy steel chain, wire rope, metal mesh, synthetic rope, synthetic round slings, and synthetic web except as described above. If Center personnel design an assembly consisting of lifting hardware produced by one or more manufacturers and build that hardware assembly, the manufacturer of each piece of hardware remains its manufacturer for purposes of determining testing.

14.5.2.3 For lifting interfaces such as eyebolts, D-rings, and lifting lugs permanently attached to the load (i.e., they will not be removed from the load prior to its use), analysis may be substituted to verify the integrity of the interface, subject to LDEM approval. To exercise this option, the following conditions shall be met:

a. The responsible organization provides documented rationale to the LDEM.

> *Note: Rationale should include design and construction information, inspections, operations, maintenance considerations, storage provisions, and other considerations.*

b. The LDEM determines there is no increase in risk.

14.5.2.4 For below-the-hook lifting devices in accordance with ASME B30.20, the proof load test value shall be as specified in that standard, or for components or subcomponents of the device, as specified in their respective ASME B30 standards, or as recommended by the designer with concurrence from the LDEM.

14.5.3 Periodic Load Test

In accordance with section 4.5, performance of a proof load test satisfies the periodic load test requirement for slings and rigging hardware.

14.5.3.1 The periodic load test for slings, rigging hardware, and below-the-hook lifting devices shall be conducted with a dummy load equal to 1.00 to 1.05 times the sling/rigging hardware/device rated capacity.

14.5.3.2 Slings and below-the-hook lifting devices shall be load tested at least once every four years unless designated as a non-load test sling or below-the-hook lifting device.

14.5.3.3 Slings and below-the-hook lifting devices shall be load tested within one year prior to use for a critical lift unless designated as a non-load test sling or below-the-hook lifting device.

14.5.3.4 Rigging hardware shall be load tested within two years prior to use for a critical lift unless one of the following applies:

a. It is designated as non-load test rigging hardware.

b. It is a lifting interface that is permanently attached to the load such as an eyebolt, D-ring, or lifting lug.

14.5.3.5 Periodic load test intervals may be extended by no more than 90 days from the original expiration date due to programmatic or institutional needs, subject to LDEM approval. To extend the periodic load test interval, the following conditions shall be met:

a. The responsible organization provides documented rationale to the LDEM.

b. The LDEM determines there is no increase in risk.

14.5.4 Non-Load Test Slings, Rigging Hardware, and Below-the-Hook Lifting Devices

14.5.4.1 Slings, rigging hardware, and below-the-hook lifting devices may be designated as non-load test slings/rigging hardware/below-the-hook lifting devices due to considerations such as usage, inspection and testing history, and potential for test induced damage, subject to LDEM approval. Non-load test slings/rigging hardware/below-the-hook lifting devices are not subject to periodic load testing requirements. To exercise this option, the following conditions shall be met:

   a. The responsible organization provides documented rationale to the LDEM.

   Note: Rationale may include design and construction information, inspections, operations, maintenance history, storage provisions, or other considerations.

   b. The LDEM determines there is no increase in risk.

14.5.4.2 The responsible organization shall label and tag non-load test slings/rigging hardware in accordance with paragraph 14.9.2 and paragraph 14.9.3.

14.6 Inspection

14.6.1 As stated in paragraph 4.1.3, inspections must comply with the applicable OSHA regulations, the VCS specified at the beginning of this chapter, the requirements in this section and section 4.6, and be based upon manufacturer recommendations.

14.6.2 Prior to use each day, the following inspections shall be performed on below-the-hook lifting devices:

   a. Check for defects such as cracks, deformations, gouges, galling, kinks, crushed areas, and corrosion.

   b. Check for proper configuration.

14.6.3 All slings, rigging hardware, and below-the-hook lifting devices rejected during inspection shall be marked and separated from accepted slings/rigging hardware/below-the-hook lifting devices.

   Note: Refer to section 14.8 for rejected sling and rigging hardware disposition.

14.7 Operation

14.7.1 As stated in paragraph 4.1.3, operations must comply with the applicable OSHA regulations, the VCS specified at the beginning of this chapter, the requirements in this section and section 4.7, and be based upon manufacturer recommendations.

14.7.2 Synthetic rope slings shall not be used for critical lifts.

Note: This requirement only applies to synthetic rope slings. It does not apply to other slings made of synthetic fibers such as synthetic round slings and synthetic web slings.

14.7.3 Slings, rigging hardware, and below-the-hook lifting devices shall not be loaded beyond the rated load except for required testing performed as outlined in section 14.5.

14.8 Maintenance

14.8.1 As stated in paragraph 4.1.3, maintenance must comply with the applicable OSHA regulations, the VCS specified at the beginning of this chapter, the requirements in this section and section 4.8, and be based upon manufacturer recommendations.

14.8.2 A designated person shall perform an engineering assessment on rejected slings, rigging hardware, and below-the-hook lifting devices to determine whether the sling and rigging hardware is repairable.

14.8.3 Non-repairable slings and rigging hardware shall be destroyed as soon as possible to avoid unintentional use.

14.9 Labeling and Tagging

14.9.1 As stated in paragraph 4.1.3, labeling and tagging must comply with the applicable OSHA regulations, the VCS specified at the beginning of this chapter, the requirements in this section and section 4.9, and be based upon manufacturer recommendations.

14.9.2 Non-load test slings, rigging hardware, and below-the-hook lifting devices shall be marked conspicuously as such.

14.9.3 Following each periodic inspection of non-load test sling/rigging hardware/below-the-hook lifting devices, a durable tag shall be affixed to the sling/rigging hardware/below-the-hook lifting device stating the next required periodic inspection date or inspection expiration date.

14.9.4 When slings and below-the-hook lifting devices are composed of an assembly of more than one sling or rigging hardware components, all load bearing sling and rigging hardware components shall be traceable to the assembly.
> *Note: This may be accomplished by clearly marking/coding or tethering all components of the assembly, through configuration control, or other procedures.*

14.10 Records

As stated in paragraph 4.1.3, record generation and retention must comply with the applicable OSHA regulations, the VCS specified at the beginning of this chapter, the requirements in section 4.10, and be based upon manufacturer recommendations. There are no additional requirements specific to slings and rigging hardware in this section.

14.11 Personnel Training and Licensing

In accordance with section 4.11, personnel training and licensing must comply with the applicable OSHA regulations, the requirements in this standard, and be based upon VCS and manufacturer recommendations. There are no additional requirements specific to slings and rigging hardware in this section.

APPENDIX A: NATIONAL AERONAUTICS AND SPACE ADMINISTRATION ALTERNATE STANDARD FOR SUSPENDED LOAD OPERATIONS

Note: In the following appendix, the term "will" is used rather than the term "shall" to indicate mandatory requirements. Terminology is not being updated because OSHA has approved the text with imperatives indicated by "will" as written.

U.S. Department of Labor

Assistant Secretary for
Occupational Safety and Health
Washington, D.C. 20210

MAR 28 2002

Mr. James D. Lloyd
Director, Safety and Risk Management Division
Office of Safety and Mission Assurance
National Aeronautics and Space Administration
Headquarters Code QS
Washington, D.C. 20546-0001

Dear Mr. Lloyd:

Thank you for your December 20, 2001 letter to the Occupational Safety and Health Administration (OSHA). We apologize for the delay in our response. Due to the October 2001 closing of the Brentwood postal facility in Washington, D.C. and the subsequent sanitizing treatment of the mail, your correspondence was not received until February 4, 2002. A letter dated February 25, 2002 amended your December 20 letter.

You requested that OSHA review and revalidate the revised version of the National Aeronautics and Space Administration's (NASA's) Alternate Standard for Suspended Load Operations. OSHA originally approved this alternate standard on July 15, 1991, in accordance with the requirements of 29 CFR 1960.17. (See enclosed letter.) The approval was based on OSHA's determination that the alternate standard provides equivalent protection as would compliance with the following standards in specifically identified operations:

- 1910.179(n)(3)(vi) The employer shall require that the operator avoid carrying loads over people.

- 1910.180(h)(3)(vi) The operator should avoid carrying loads over people.

- 1910.180(h)(4)(ii) No person should be permitted to pass under a load on the hook.

OSHA Alternate Standard on Suspended Loads Revalidation Letter

The alternate standard is currently a part of NASA's Safety Standard for Lifting Devices and Equipment (NSS/GO-1740.9B). NASA is in the approval process for updating and issuing the safety standard in a new format as the Standard for Lifting Devices and Equipment (NASA-STD-8719.9). NASA intends to include a revised version of the alternate standard on suspended loads as part of NASA-STD-8719.9. The minor revisions NASA proposes to the existing alternate standard are:

- Renumbering of the alternate standard paragraphs to be consistent with the numbering conventions of NASA Technical Standards.

- Changing the references from NSS/GO-1740.9B to NASA-STD-8719.9 to reflect the release of the new NASA Technical Standard.

The revisions you propose to the existing alternate standard were reviewed. It has been determined that they will not affect the existing alternate standard as they are administrative in nature.

Thank you for your interest in occupational safety and health. If you have any questions, please do not hesitate to contact Thomas K. Marple, Director, Office of Federal Agency Programs, at (202) 693-2122.

Sincerely,

John L. Henshaw

Enclosure

OSHA Alternate Standard on Suspended Loads Revalidation Letter

U.S. Department of Labor
Assistant Secretary for Occupational Safety and Health
Washington, D.C. 20210

DEC 15 1991

Mr. George A. Rodney
Associate Administrator for
  Safety and Mission Quality
National Aeronautics and Space
  Administration
600 Independence Avenue, S.W.
Washington, D.C. 20546

Dear Mr. Rodney:

The Occupational Safety and Health Administration (OSHA) has completed its review of the proposed alternate standard on suspended loads, as required in 29 CFR 1960.17. With this letter, we want to inform you that we have approved the standard. This approval is based on our determination that the alternate standard provides equivalent protection as would compliance with the following standards in specifically identified operations:

- 1910.179(n)(3)(vi) The employer shall require that the operator avoid carrying loads over people.

- 1910.180(h)(3)(vi) The operator should avoid carrying loads over people.

- 1910.180(h)(4)(ii) No person should be permitted to pass under a load on the hook.

One of the OSHA reviewers stated that this standard, ". . . appears to be a very comprehensive approach to a finite task and requires significant amounts of safety management from the preliminary hazard analysis through completion of the lift." It is essential, however, that management ensure that this level of safety management effort continues to effectively protect the exposed employees.

We appreciate the cooperation provided my staff in the many discussions on this alternate standard. Your interest and support for the safety and health of Federal employees is greatly appreciated.

Sincerely,

Gerard F. Scannell
Assistant Secretary

OSHA Alternate Standard on Suspended Loads Validation Letter

NASA-STD-8719.9B – 2018-10-25

NATIONAL AERONAUTICS AND SPACE ADMINISTRATION ALTERNATE STANDARD FOR SUSPENDED LOAD OPERATIONS

A.1 This standard applies to specifically identified operations controlled by the National Aeronautics and Space Administration (NASA) involving both civil service and contractor employees. The standard is an alternate to Code of Federal Regulations 29 CFR 1910.179(n)(3)(vi), 29 CFR 1910.180(h)(3)(vi), and 29 CFR 1910.180(h)(4)(ii). NASA Safety is responsible for its implementation and enforcement.

A.2 As an alternative standard developed pursuant to Section 1-201(d) of Executive Order 12196 and 29 CFR 1960.17, it applies only to NASA employees. The Occupational Safety and Health Administration (OSHA) will inspect the working conditions of NASA employees performing these specified operations for compliance with these alternate standard requirements. Although OSHA cannot inspect private sector employees working in the same operation with NASA employees for compliance with the alternate standard, it will fully consider the equivalent safeguards specified in this standard for both NASA and contractor employees as the basis for a de minimis violation which is recorded, but not issued.

A.3 Suspended Load Operation Definition. An operation is considered a suspended load operation and subject to the requirements of this standard if it meets all three of the following criteria:

  A.3.1 The operation involves the use of a crane or hoist that supports the weight of a suspended load. (This excludes operations where the load is secured in a holding fixture or on substantial blocks supporting the entire load even though the crane/hoist hook may still be attached.) No distinction is made between a static load and a dynamic load. Rigging, i.e., slings, Hydra Sets, lifting fixtures, shackles, straps, when attached to the hook, is considered part of the load.

  A.3.2 Personnel involved in the operation have any part of the body directly beneath the suspended load. (This excludes operations where employees have their hands on the sides of a load, i.e., to guide the load.)

  A.3.3 In the event of a crane/hoist failure, as the load drops it could contact personnel working directly beneath it, with injury or death as a possible result. (This excludes operations where employees have their hands only partially under a load such that a crane or hoist device failure would push their hands out of the way not resulting in injury. This also excludes situations where the falling load would come to rest on hardware that is not suspended before an employee could be injured.)

A.4 Requirements. It is recognized that cranes and hoists do not generally meet the support requirements of a system that would allow personnel to work beneath a suspended load. NASA's first hazard avoidance protocol is to design hazards out of the system or operation. Accordingly, it is NASA's intent and goal that all future systems, hardware, and equipment be engineered, designed, installed, and operated to prevent exposing employees to working under loads suspended from cranes and hoists. Due to the uniqueness of NASA activities and the limitations imposed when using present systems, hardware, equipment, and facilities, suspended load operations may be permitted only under specifically approved and controlled conditions. No suspended load operation shall be performed unless all (15) of the following special requirements are met:

A.4.1　All suspended load operations will be approved by the Center/facility NASA Director of Safety based upon a detailed engineering hazards analysis of the operation. The hazards analysis will be prepared by the responsible safety organization and coordinated through appropriate engineering and design offices. The analysis documentation will include the following:

  a.　A justification why the operation cannot be conducted without personnel beneath the load. Feasible procedure/design options will be investigated to determine if the work can be accomplished without personnel working under a load suspended from a crane/hoist.

  b.　Details of the precautions taken to protect personnel should the load drop. Secondary support systems, i.e., equipment designed to assume support of (catch) the load preventing injury to personnel should the crane/hoist fail, shall be evaluated and used whenever feasible. Secondary support systems will be constructed with a minimum safety factor of 2 to yield.

  c.　The maximum number of exposed personnel allowed. Steps shall be taken to limit the number of personnel working under a load suspended from a crane/hoist. Only those essential personnel absolutely necessary to perform the operation will be allowed to work in the safety controlled area.

  d.　The time of exposure. Steps shall be taken to ensure that personnel do not remain under the load any longer than necessary to complete the work.

A.4.2　Each operation will be reviewed on a case-by-case basis.

A.4.3　Only those suspended load operations approved by the Center/facility NASA Director of Safety will be permitted, subject to this standard. A list of approved suspended load operations will be maintained by NASA Safety and made available to OSHA personnel upon request.

A.4.4　The operational procedures document (e.g., Operations and Maintenance Instruction, Technical Operating Procedure, Work Authorization Document) will be revised to specify the necessary additional requirements identified by the hazard analysis discussed in paragraph A.4.1. The procedures will be available on site for inspection during the operation.

A.4.5　During a suspended load operation, if a new procedure not covered by the original analysis is deemed necessary due to unusual or unforeseen circumstances, the NASA Center/facility Safety Office will be consulted and must approve and document the procedure before operations continue. Safety will coordinate with Operations, Engineering, and other organizations as appropriate. If the new procedure is to be performed on a regular basis, a detailed hazards analysis and approval as outlined in paragraph A.4.1 are required.

A.4.6　The crane/hoist shall be designed, tested, inspected, maintained, and operated in accordance with the NASA Standard for Lifting Devices and Equipment (NASA-STD-8719.9).

Test, inspection, and maintenance procedures will be developed and approved by qualified, responsible NASA engineers. Qualified specialists will perform the procedures and resolve noted discrepancies. NASA Quality Assurance will perform an independent annual inspection of all cranes/hoists involved in suspended load operations. The results of the annual inspections will be maintained and made available to OSHA personnel upon request.

A.4.7 Each crane/hoist involved in suspended load operations shall undergo a Failure Modes and Effects Analysis (FMEA) that shall be approved by the Center/facility NASA Director of Safety. The FMEA will determine Single Failure Points (SFP), assessing all critical mechanical functional components and support systems in the drive trains and critical electrical components.

    a. For those cranes/hoists identified as having no SFP whose failure would result in dropping the load, the total weight of the suspended load shall not exceed the device's rated load.

    b. For those cranes/hoists identified as having a SFP whose failure would result in dropping the load, use of that device for suspended load operations must be approved by NASA Headquarters. Complete documentation on the suspended load operation, including the hazards analysis outlined in paragraph A.4.1 and the FMEA described above, will be forwarded to NASA Headquarters for evaluation. Approval will be given based upon detailed analysis of the potential hazards and rationale for acceptance. Such cases will never exceed the device's rated load. OSHA shall be notified when NASA Headquarters approves using any crane/hoist identified as having a SFP whose failure would result in dropping the load.

A.4.8 Before lifting the load involved in a suspended load operation, the crane/hoist will undergo a visual inspection (without major disassembly) of components instrumental in assuring that the load will not be dropped (e.g., primary and secondary brake systems, hydraulics, mechanical linkages, and wire rope per NASA-STD-8719.9). Noted discrepancies will be resolved before the operation continues. This pre-lift inspection will be in addition to the inspections required in 1910.179(j) and 180(d).

A.4.9 A trained and licensed operator (certified per NASA-STD-8719.9) shall remain at the crane/hoist controls while personnel are under the load.

A.4.10 Safety controlled areas shall be established with appropriate barriers (rope, cones, etc.). All nonessential personnel shall be required to remain behind the barriers.

A.4.11 Prior to the suspended load operation, a meeting with the crane/hoist operator(s), signal person(s), person(s) who will work under the load, and the person responsible for the task shall be held to plan and review the approved operational procedures that will be followed, including procedures for entering and leaving the safety controlled area.

A.4.12 Communications (voice, radio, hard wired, or visual) between the operator(s), signal person(s), and the person(s) working under the load shall be maintained. Upon communication loss, operations shall stop immediately, personnel shall clear the hazardous area, and the load shall be safed. Operations shall not continue until communications are restored.

A.4.13 Personnel working beneath the load shall remain in continuous sight of the operator(s) and/or the signal person(s).

A.4.14 NASA shall conduct periodic reviews to ensure the continued safety of the procedures. As a minimum, NASA will annually evaluate the implementation of this procedure at each Center with operations on the suspended load list.

A.4.15 A list of approved suspended load operations, list of cranes/hoists used for suspended load operations, and copies of the associated hazards analyses will be provided to the OSHA Office of Federal Agency Programs via NASA Headquarters for distribution to the appropriate regional and area OSHA offices. (NASA Headquarters, in conjunction with OSHA, will develop a format for transmittal of this information.) Quarterly updates to the documentation will be provided as needed.

APPENDIX B: SUMMARY OF CRITICAL LIFT REQUIREMENTS

Critical lifts are those lifts where failure/loss of control presents an elevated risk of serious injury, loss of life, or loss of one-of-a-kind articles; high dollar items or major facility components the loss of which would have serious programmatic or institutional impact. Lifts of high-value spacecraft are usually classified as critical lifts, while lifts of small, improvised mini-satellites, for example, most likely would not be. Lifting and movement of flight hardware components packaged per applicable shipment specifications are typically not classified as critical lifts.

Critical LDE is the equipment used to perform critical lifts.

Requirements that apply exclusively to critical lifts and critical LDE are summarized in this appendix. Some requirements are paraphrased in this appendix to provide context without unnecessarily quoting a large block of text. Paraphrased passages are contained in brackets.

Chapters 1 through 3 -- None

Chapter 4: General LDE Requirements

- The responsible organization shall follow a documented process that seeks input from the appropriate stakeholders (such as facility, program, operations, and safety) and the LDEM to classify lifts as critical or noncritical. (4.2.2)

- The LDEM shall have the authority to reclassify noncritical lifts as critical based upon safety, facility, or other concerns beyond normal lifting operations. (4.2.3)

- An operation shall be classified as a critical lift when failure/loss of control presents an elevated risk of serious injury, loss of life, or loss of one-of-a-kind articles, high dollar items or major facility components whose loss would have serious programmatic or institutional impact. (4.2.4)

  Note: Lifts of high-value spacecraft are usually classified as critical lifts, while lifts of small, improvised mini satellites, for example, most likely would not be. Lifting and movement of flight hardware components packaged per applicable shipment specifications are typically not classified as critical lifts.

- An operation may be classified as a noncritical lift if it does not meet critical lift criteria. (4.2.5)

  Note: Noncritical lifts typically involve routine lifting operations and are governed by standard industry rules and practices except as supplemented with unique NASA testing, operations, maintenance, inspection, and personnel licensing requirements contained in this standard.

- A recognized safety hazard analysis shall be performed on critical or custom-built LDE (subject to documented LDEM approval, hooks, rigging hardware, slings, below-the-hook lifting devices, and attachments for industrial trucks may be excluded for cases in which there is no potential for load instability). (4.3.1)

- The safety hazard analysis shall, as a minimum, identify potential sources of danger and recommend resolutions for those conditions that could cause loss of life, personal injury, and loss of or damage to the LDE, facility, or load. (4.3.2)

- When critical or custom-built LDE is designed or procured, the responsible organization shall notify the LDEM and provide the LDEM with the information necessary for review and approval (subject to documented LDEM approval, hooks, rigging hardware, and slings may be excluded from this requirement). (4.4.2)

- Specific written procedures shall be prepared and followed for critical lifts. (4.7.9)

- Critical LDE shall be marked conspicuously as such. (4.9.3)

    Note: Hooks that are part of critical LDE or attachments that are permanently mounted on industrial trucks do not need separate marking.

Chapter 5: Overhead Cranes

- Cranes used for critical lifts shall have one of the following: (5.4.1.1)

    a. Two holding brakes, each capable of bringing a rated load to zero speed and holding it.

    Note: A load brake may be considered a second holding brake provided it is capable of bringing a rated load to zero speed and holding it.

    b. A single holding brake in combination with a motor drive that automatically monitors brake functionality and motor torque.

- When used for critical lifts, speed reduction from the motor to the drum on the hoist should be achieved by using gears enclosed in a gear case. If open gears are required, they shall be guarded, with provision for lubrication and inspection. (5.4.1.4)

- Cranes used for critical lifts shall be equipped with dual upper limit switches. (5.4.2.5)

- For critical lift electric cranes, the limit switches shall meet the following: (5.4.2.6)

    a. The initial upper limit switch precludes movement in the raise direction when the limit is reached.

    Note: Movement in the "lower" direction need not be inhibited in association with the initial upper limit switch function.

    b. The final upper limit switch is wired into the mainline circuit, hoist power circuit, main contactor control circuit, or hoist power contactor control circuit, such that all crane motion or all hoist motion is precluded when the limit is reached.

    c. After a final upper limit switch has been activated, movement of the load requires action (resetting) at the final upper limit switch level.

Note: The crane design should include a means of detecting limit switch failure and allow for safe inspection and repair. For example, a system may be equipped with two different colored annunciator lights, one for each limit switch. A reset button may be included so when a final upper limit switch is tripped, the load can be lowered immediately. The reset button should be secured to prevent unauthorized or unintended use.

d. The initial upper limit switch is adjusted sufficiently low to preclude inadvertent actuation of the final upper limit switch if the hoist actuates the initial upper limit switch at full speed with no load. Similarly, the final upper limit switch is adjusted sufficiently low to ensure the hoist will not two-block (or otherwise damage wire rope) if the hoist actuates the final upper limit switch at full speed with no load.

Note: This requirement effectively lowers the usable hook height of the hoist.

- For cranes used for critical lifts, a lower limit switch shall be provided to ensure no less than two wraps remain on the drum. (5.4.2.7)

    Note: Movement in the "raise" direction need not be inhibited in association with the lower limit switch function.

- Critical lift cranes should have a fail-safe control system such that a single failure does not cause the crane to operate at a speed faster than commanded or in a direction other than commanded. (5.4.2.8)

    Note: A failure that stops the crane and sets the brakes, or causes the crane to operate in a speed slower than commanded without disabling the stop function is acceptable.

- A periodic load test shall have been performed on a crane within one year prior to its use for a critical lift. (5.5.2.2)

Chapter 6: Mobile Cranes and Derricks

- A periodic load test shall have been performed on a mobile crane/derrick within one year prior to its use for a critical lift. (6.5.2.3)
- A periodic load test shall be performed after each boom change (when boom disassembly/assembly is required) if the mobile crane/derrick is to be used for critical lifts. (6.5.2.5)

Chapter 7: Hoists and Winches

- For powered hoists and winches: (7.4.1.1)

a. Except as provided in paragraph 7.4.1.1.b, hoists/winches used for critical lifts shall have one of the following:

(1) Two holding brakes, each capable of bringing a rated load to zero speed and holding it.

Note: A load brake may be considered a second holding brake provided it is capable of bringing a rated load to zero speed and holding it.

(2) A single holding brake in combination with a motor drive that automatically monitors brake functionality and motor torque.

b. Subject to LDEM approval, a hoist/winch with a single holding brake may be used for critical lifts when hoists/winches compliant with paragraphs 7.4.1.1.a(1) and (2) are not commercially available. To exercise this option, the following conditions shall be met:

(1) The responsible organization provides documented rationale to the LDEM.

Note: Rationale may include design and construction information, inspections, operations, maintenance and storage provisions, or other considerations.

(2) The LDEM determines there is no increase in risk.

c. When used for critical lifts, speed reduction from the motor to the drum on the hoist should be achieved by using gears enclosed in a gear case. If open gears are required, they shall be guarded with a provision for lubrication and inspection.

- Over-travel Protection for Powered Hoist and Winches (7.4.2)

7.4.2.1. Except as provided in paragraph 7.4.2.5 and 7.4.2.6, electric and air-powered hoists and winches used for critical lifts shall be equipped with dual upper limit switches.

7.4.2.2 Except as provided in paragraph 7.4.2.5, electric and air-powered hoists and winches used for critical lifts shall be equipped with a lower limit switch to ensure no less than two wraps remain on the drum.

Note: Movement in the "raise" direction need not be inhibited in association with the lower limit switch function.

7.4.2.3. For air-powered hoists and winches equipped with dual upper limit switches, the final upper limit switch shall exhaust air from the hoist or winch, set the brakes, and require reset at the upper limit switch level.

7.4.2.4. Electric hoists and winches equipped with dual upper limit switches shall meet the following:

a. The initial upper limit switch precludes movement in the raise direction when the limit is reached.

Note: Movement in the "lower" direction need not be inhibited in association with the initial upper limit switch function.

b. The final upper limit switch is wired into the mainline circuit, hoist/winch power circuit, main contactor control circuit, or hoist/winch power contactor control circuit, such that all hoist/winch motion is precluded when the limit is reached.

c. After a final upper limit switch has been activated, movement of the load requires action (resetting) at the final upper limit switch level.

Note: The hoist design should include a means of detecting limit switch failure and allow for safe inspection and repair. For example, a system may be equipped with two different colored annunciator lights, one for each limit switch. A reset button may be included so when a final upper limit switch is tripped, the load can be lowered immediately. The reset button should be secured to prevent unauthorized or unintended use.

d. The initial upper limit switch is adjusted sufficiently low to preclude inadvertent actuation of the final upper limit switch if the hoist actuates the initial switch at full speed with no load. Similarly, the final upper limit is adjusted sufficiently low to ensure the hoist or winch will not two-block (or otherwise damage wire rope) if the hoist or winch actuates the final switch at full speed with no load.

Note: This requirement effectively lowers the usable hook height of the hoist.

7.4.2.5 Subject to LDEM approval, a hoist/winch with a single upper limit and no lower limit switch may be used for critical lifts when hoists/winches compliant with paragraphs 7.4.2.1 and 7.4.2.2 are not commercially available. To exercise this option, the following conditions shall be met:

a. The responsible organization provides documented rationale to the LDEM.

Note: Rationale may include design and construction information, inspections, operations, maintenance and storage provisions, or other considerations.

b. The LDEM determines there is no increase in risk.

7.4.2.6 Air-powered chain hoists may use a travel-limiting clutch in place of the final upper limit switch.

- A periodic load test shall have been performed on a hoist/winch within one year prior to its use for a critical lift (7.5.2.3).

Chapter 8: Hoist-supported Personnel Lifting Devices

- Lifting of personnel using a hoist-supported personnel lifting device shall be classified as a critical lift. (8.2.2)
- A recognized safety hazard analysis shall be performed on hoist-supported personnel lifting devices. (8.3.2)

Chapter 9: Mobile Aerial Platforms -- None
Chapter 10: High Lift Industrial Trucks

- A periodic load test shall have been performed on an industrial truck and its attachments within one year prior to their use for a critical lift. (10.5.2.2)

Chapter 11: Load Positioning and Load Measuring Devices

- A periodic load test shall have been performed on load positioning and load measuring devices within one year prior to their use for a critical lift. (11.5.2.3)

Chapter 12: Jacks

- Chapter 12 applies to jacks used for critical lifts.

Chapter 13: Hooks

- Hooks on overhead cranes; mobile cranes; and critical lift hoists; shall be given a surface NDT immediately after all proof load and periodic load tests prior to further use of the hook. (13.6.3)

Chapter 14: Slings and Rigging Hardware

- Slings and below-the-hook lifting devices shall be load tested within one year prior to use for a critical lift, unless designated as a non-load test sling or below-the-hook lifting device. (14.5.3.3)

- [Rigging hardware shall be load tested within two years prior to use for a critical lift unless designated as non-load test rigging hardware or permanently attached to the load.] (14.5.3.4)

- Synthetic rope slings shall not be used for critical lifts. (14.7.2)

    Note: This requirement only applies to synthetic rope slings. It does not apply to other slings made of synthetic fibers such as synthetic round slings and synthetic web slings.

APPENDIX C: LDEM ROLES, APPROVALS, AND SPECIAL PERMISSIONS

LDEM responsibilities, approval, and special permissions are summarized in this appendix. The location of the requirement is noted in parentheses after the requirement.

Some requirements were paraphrased in this appendix to provide context without unnecessarily quoting a large block of text. Paraphrased passages are contained in brackets.

Chapter 1: Scope

- Rented or leased LDE used for non-critical lifts may be exempted from this standard by the written decision of the contracting officer, the responsible NASA installation/program safety office, and the LDEM, based on an assessment of associated risk. (1.2.2)
- The need for compliance with this standard at contractor installations performing NASA work should be evaluated and made a contractual requirement where deemed necessary by the contracting officer, the responsible NASA installation/program safety office, and the LDEM. (1.2.3)
- The LDEM shall have the authority to interpret this standard. (1.2.4)
- The LDEM shall have the authority to approve, disapprove, and levy requirements for the use of LDE not covered by paragraph 1.2.1 due to safety concerns or hazards presented by a particular application. (1.2.5)
- [The LDEM shall participate as a member of the NASA LDE Committee, chair the Center/Facility LDE Committee, and serve as the focal point for implementation, clarification, and enforcement of NASA-STD-8719.] (1.3.6)
- [Coordinate with the responsible organizations at the Center/Facility to implement this standard.] (1.3.7)
- In case of questions regarding conflicting requirements, the applicability of this standard, or to request a clarification, contact the LDEM. (1.6.3)

Chapter 2: Applicable Documents

- Compliance with the following [non-government publications] is required as specified herein. Equivalent standards may be substituted with approval from the LDEM. (2.3)

Chapter 3: Definitions and Acronyms -- None

Chapter 4: General LDE Requirements

- LDEM approval shall be obtained for any tailoring of manufacturer recommendations. (4.1.4)

- The responsible organization shall follow a documented process that seeks input from the appropriate stakeholders (such as facility, program, operations, and safety) and the LDEM to classify lifts as critical or noncritical. (4.2.2)

- The LDEM shall have the authority to reclassify noncritical lifts as critical based upon safety, facility, or other concerns beyond normal lifting operations. (4.2.3)

- A recognized safety hazard analysis shall be performed on critical or custom-built LDE (subject to documented LDEM approval, hooks, rigging hardware, slings, below-the-hook lifting devices, and attachments for industrial trucks may be excluded for cases in which there is no potential for load instability). (4.3.1)

- When critical or custom-built LDE is designed or procured, the responsible organization shall notify the LDEM and provide the necessary information for review and approval (subject to documented LDEM approval, hooks, rigging hardware, and slings may be excluded from this requirement). (4.4.2)

- LDEM approval shall be obtained for any modifications to LDE. (4.8.4)

    Note: Replacement in kind is not considered a modification and does not require LDEM approval.

- Personnel operating LDE shall be appropriately trained and licensed. (4.11.1.1)

    Note: LDE operators must be appropriately trained. This standard does not require a license to operate manually operated hoists and winches, personnel access platform hoists/winches, manually propelled mobile aerial platforms (e.g., access stand/stairs), manually propelled industrial trucks, manually operated load positioning devices, load measuring devices, and jacks, but additional licensing may be required by Center policy or the LDEM.

- Licensing organizations and the LDEM shall reserve the right to suspend or revoke licenses for reasons such as negligence, violations of requirements, or failure to meet medical standards. (4.11.2.3)

- The LDEM shall review the personnel licensing program at least annually to ensure the contents, training material, testing, and examination elements are up-to-date with current methods and techniques and any "lessons-learned" are adequately addressed. (4.11.2.8)

Chapter 5: Overhead Cranes

- [Approve/disapprove equivalent VCS.] (5.1.2)

- Cast iron components shall not be used in the hoist load path unless approved by the LDEM and the responsible organization. (5.4.1.6)

    Note: The material properties of cast iron allow catastrophic failure (brittle fracture), and it should not be considered as reliable as steel or cast steel. The engineer should consider this when selecting equipment and avoid the use of load bearing cast iron materials where possible.

- [Approve/disapprove requests to extend the periodic load test interval by no more than 90 days due to programmatic or institutional need.] (5.5.2.3)
- Consult with the LDEM regarding appropriate range of travel [when testing hoist, bridge, and trolley under load]. (5.5.2.4)
- [Approve/disapprove cranes for use in]…load testing items such as slings, platforms, and lifting fixtures, or to relieve a portion of the weight of a constrained load. [5.7.3]
- [Approve/disapprove alternative handling procedures other than E-Stops for use when the operator's view is restricted/obstructed.] [5.7.6]
- [Approve/disapprove requests to omit crane directional markings.] [5.9.2]

Chapter 6: Mobile Cranes and Derricks

- [Approve/disapprove equivalent VCS.] (6.1.2)
- [Approve/disapprove requests to exceed 1.00 times the rated capacity of mobile cranes/derricks for proof load testing.] (6.5.1.2)
- [Approve/disapprove requests to extend the periodic load test interval by no more than 90 days due to programmatic or institutional need.] [6.5.2.4]
- [Approve/disapprove requests to use other methods to perform periodic load tests.] (6.5.2.7)
- [Approve/disapprove cranes/derricks for use in]…load testing items such as slings, platforms, and lifting fixtures or to relieve a portion of the weight of a constrained load. (6.7.3)

Chapter 7: Hoists and Winches

- [Approve/disapprove equivalent VCS] (7.1.2)
- [Approve/disapprove requests to use hoists/winches with a single holding brake for critical lifts due to inability to commercially source hoists/winches with either two holding brakes or a drive that monitors brake and motor functionality.] (7.4.1.1)
- Cast iron components shall not be used in the hoist or winch load path unless approved by the LDEM and the responsible organization. (7.4.1.3)

    Note: The material properties of cast iron allow catastrophic failure (brittle fracture), and it should not be considered as reliable as steel or cast steel. The engineer should consider this when selecting equipment and avoid the use of load bearing cast iron materials where possible.

- [Approve/disapprove requests to use hoists/winches with a single upper limit switch and no lower limit switch for critical lifts due to inability to commercially source hoists/winches with two upper limit switches and a lower limit switch.] (7.4.2.5)
- [Approve/disapprove proof load testing method for replacement in kind of powered hoists/winches on existing mounting structures.] (7.5.1.3)

- [Approve/disapprove requests to extend the periodic load test interval by no more than 90 days due to programmatic or institutional need.] (7.5.2.4)
- Consult the LDEM regarding appropriate range of travel [when testing hoist/winch under load]. (7.5.2.5)
- [The LDEM may approve alternative methods to test the holding brakes as part of periodic load test for Hoists and Winches.] (7.5.2.5)
- [The LDEM may approve handling procedures that minimize risk when Hoist/Winch operator's view is obstructed.] (7.7.3)
- [LDEM may approve a hoist/winch for load testing items such as slings, platforms, and lifting fixtures or to relieve a portion of the weight of a constrained load.] (7.7.8)

Chapter 8: Hoist-Supported Personnel Lifting Devices

- Hoist-Supported personnel lifting devices shall have at least one of the following: (8.4.2)

  a. Two independent support systems consisting of two separate hoists such that the failure of one hoist, its reeving system, or other component will not cause the stability of the personnel lifting device to be lost or prohibit its movement to a safe location.

  b. A single support system with two or more holding brakes and additional factors of safety for the hoist and other load bearing components as approved by the LDEM.

  c. Other methods/attributes as approved by the LDEM.

- [Approve/disapprove requests to extend the periodic load test interval by no more than 90 days due to programmatic or institutional need.] (8.5.2.3)
- Consult the LDEM regarding appropriate range of travel [when testing hoist under load]. (8.5.2.4)
- [The LDEM may approve alternative methods to test the holding brakes as part of periodic load test for Hoist-supported Personnel Lifting Devices.] (8.5.2.4)

Chapter 9: Mobile Aerial Platforms

- [Approve/disapprove equivalent VCS.] (9.1.2)
- [Approve/disapprove requests to extend the periodic load test interval by no more than 90 days due to programmatic or institutional need.] (9.5.2.3)

Chapter 10: High Lift Industrial Trucks

- [Approve/disapprove equivalent VCS.] (10.1.2)
- [Approve/disapprove requests to extend the periodic load test interval by no more than 90 days due to programmatic or institutional need.] (10.5.2.3)

Chapter 11: Load Positioning and Load Measuring Devices

- [Approve/disapprove equivalent VCS.] (11.1.2)

- For load positioning devices, the proof load test shall consist of holding a dummy load of 1.20 to 1.25 times the rated capacity or as recommended by the designer with concurrence from the LDEM. (11.5.2.6)
- [Approve/disapprove requests to extend the periodic load test interval by no more than 90 days due to programmatic or institutional need.] (11.5.2.4)

Chapter 12: Jacks

- [Approve/disapprove equivalent VCS] (12.1.2)
- [The requirements in chapter 12 may be applied to non-critical jacks at the discretion of the LDEM.] (12.1.1)
- [Approve/disapprove requests to extend the periodic load test interval by no more than 90 days due to programmatic or institutional need.] (12.5.2.3)

Chapter 13: Hooks

- [Approve/disapprove equivalent VCS.] (13.1.2)
- [Approve/disapprove requests to extend the periodic surface NDT interval by no more than 4 years due to programmatic or institutional need.] (13.6.4)
- Volumetric NDT shall be conducted on new hooks at the discretion of the LDEM and the responsible organization. (13.6.5)

Chapter 14: Slings, Rigging Hardware, and Below-the-Hook Lifting Devices

- [Approve/disapprove equivalent VCS.] (14.1.2)
- [Approve/disapprove requests to substitute analysis to verify the integrity of lifting interfaces such as eyebolts, D-rings, and lifting lugs permanently attached to the load in lieu of a proof load test.] (14.5.2.3)
- For below-the-hook lifting devices, the proof load test value shall be 1.20 to 1.25 times the rated capacity of the sling or as recommended by the designer with concurrence from the LDEM. (14.5.2.4)
- [Approve/disapprove requests to extend the periodic load test interval by no more than 90 days due to programmatic or institutional need.] (14.5.3.5)
- [Approve/disapprove requests to designate slings, rigging hardware, and below-the-hook lifting devices as non-load test slings/rigging hardware/below-the-hook lifting devices.] (14.5.4.1)
- [Approve/disapprove requests to designate slings, rigging hardware, and below-the-hook lifting devices as non-load test slings/rigging hardware/below-the-hook lifting devices.] (14.5.4.1)

APPENDIX D: RECOMMENDED MINIMUM LDEM QUALIFICATIONS

Introduction:

The function of the Lifting Devices and Equipment Manager (LDEM) is to act as the Center lead and Technical Authority in managing risk to people, facilities, and the environment posed by the operation of lifting devices and equipment (LDE) as established in NASA STD 8719.9. In addition to a number of specific functions, the LDEM is to serve as the SMA Technical Authority for LDE.

The following represent the recommended minimum attributes for this position:

(1) Education and experience:

    a. A bachelor's degree in engineering from an accredited college or university and a minimum of 2 years of cumulative experience with lifting devices and equipment design, specification development, procurement, maintenance, inspection, and operations, or

    b. An associate's or bachelor's degree in engineering technology from an accredited college or university and a minimum of 5 years of cumulative experience with lifting devices and equipment design, specification development, procurement, maintenance, inspection, and operations.

(2) Knowledge requirements:

    a. Thorough understanding of:

        i. Stress analysis, including static and dynamic analysis as encountered in mechanical systems,

        ii. Statics and dynamics,

        iii. Fracture Mechanics,

        iv. Fabrication, welding, inspection, etc. techniques and processes as related to LDE construction and repair,

        v. LDE failure modes and damage mechanisms, and the NDE techniques and analyses that are used to identify precursors to those failure modes and to determine acceptable limits on operations so as to avoid them, and

        vi. Wire rope load capabilities, use, maintenance, wear, and failure mechanisms.

b. Knowledge of ASME and other LDE VCS, related Occupational Safety and Health Administration (OSHA) regulations, and NASA policies, requirements, and procedures associated with LDE.

(3) Ability to:

a. Develop and/or review and assess LDE designs, drawings, and procurements specifications and statements of work for clarity, accuracy, and compliance with applicable requirements, and provide approval/disapproval recommendations to Center management,

b. Assess and document LDE for adequate design and quality for the intended function, and ensure their fitness for service and safety for continued operation in accordance with the requirements of NASA STD 8719.9 and associated VCS and applicable regulations,

c. Advise Center management, the Contracting Officer, and other management and project management personnel on LDE issues, including budgets, procurements, certification and training requirements, operations, disposition of requests for relief from requirements, etc.,

d. Represent the Center on the Agency wide Lifting Devices and Equipment Committee (LDEC) and to the Headquarters Office of Safety and Mission Assurance, and chair the Center LDEC,

e. Develop and maintain Center LDE policies,

f. Plan and manage the Center LDE program, including developing budgets, schedules, and work plans to minimize risk associated with Center LDE operations as Center SMA Technical Authority (TA) for LDE,

g. Work effectively with all levels of Center technical, management, and operations personnel to ensure safe LDE operations through understanding and implementation of LDE requirements, including the following:

　　i. NASA and Center Standards / Directives / Processes,

　　ii. VCS,

　　iii. OSHA regulations and other applicable regulatory requirements,

h. Communicate effectively both verbally and in writing in technical and project management scenarios,

i. Develop and/or review, and approve clear, complete, and accurate specifications and statements of work for LDE procurements or modifications,

j. Physically access cranes and other LDE in the field for inspection and assessment, as needed,

k. Assess and document compliance of new and existing LDE, LDE training, and LDE operators with NASA-STD-8719.9 and applicable OSHA regulations and VCS.

(4) Be conversant with the Federal procurement process, NASA and Center standards, directives, and processes, and hazard analysis processes within six months of hire date.

www.ingramcontent.com/pod-product-compliance
Lightning Source LLC
Chambersburg PA
CBHW062329220526

45469CB00008B/2646